Copper Bioinorganic Chemistry

From Health to Bioinspired Catalysis

Copper Bioinorganic Chemistry

From Health to Bioinspired Catalysis

Editors

A Jalila Simaan • Marius Réglier

Centre Nationale de la Recherche Scientifique, France &
Aix Marseille Université, France

World Scientific

NEW JERSEY · LONDON · SINGAPORE · BEIJING · SHANGHAI · HONG KONG · TAIPEI · CHENNAI · TOKYO

Published by

World Scientific Publishing Co. Pte. Ltd.

5 Toh Tuck Link, Singapore 596224

USA office: 27 Warren Street, Suite 401-402, Hackensack, NJ 07601

UK office: 57 Shelton Street, Covent Garden, London WC2H 9HE

British Library Cataloguing-in-Publication Data
A catalogue record for this book is available from the British Library.

ISBN 978-981-126-948-6 (hardcover)
ISBN 978-981-126-949-3 (ebook for institutions)
ISBN 978-981-126-950-9 (ebook for individuals)

For any available supplementary material, please visit
https://www.worldscientific.com/worldscibooks/10.1142/13237#t=suppl

Typeset by Stallion Press
Email: enquiries@stallionpress.com

Preface

Bioinorganic chemistry is a booming field of research that has its source at the frontiers of chemistry, biochemistry, molecular biology, medicinal chemistry, all disciplines that are in perpetual mutual enrichment. While in the late 1970s bioinorganic chemistry focused mainly on metals such as iron (hemoproteins, iron/sulfur proteins …) or zinc (aminopeptidases …), over the years it has gradually developed towards other metals including copper. One reason for the initial low interest in copper biochemistry was the few known copper-containing biological systems. In the early 1980s, most studies on copper proteins concerned those on electron transfer proteins (azurin, plastocyanin, …). Another reason was probably due the fact that within the two oxidation states of copper, *i.e.* Cu(I) and Cu(II) involved in biological systems, only the Cu(II) oxidation state was accessible by spectroscopic methods commonly used in these years (UV-vis. and EPR) while Cu(I) oxidation state remained invisible. Over the past few decades, advances in chemistry, spectroscopic methods and systems biology have enabled several discoveries about the role of copper in human pathologies (e.g. Menkes, Wilson and Alzheimer diseases) as well as on the molecular functioning of copper-containing systems. This was accompanied by growing discovery of copper-containing biological systems, shedding a new light on this metal ion.

Copper is an essential micronutrient for most living beings including bacteria. This is mainly due to essential roles of catalytic Cu-centers in enzymes. However, an excess of Cu is toxic and it has thereby been used over centuries as a powerful antimicrobial agent. Bacteria developed several systems to detect Cu and protect themselves from high intracellular Cu concentration, namely via Cu-extrusion and/or Cu- storage. Chapter 1 entitled *"Ligands as a tool to tune the toxicity of Cu on bacteria: from boosting to silencing*" by Peter Faller, Marianne Ilbert *et al.* addresses this topic focusing on Cu-detoxification systems and progress in understanding the molecular mechanisms of Cu-toxicity for bacteria.

Tyrosinase is an iconic quasi-ubiquitous mono-oxygenase/oxidase studied since the early 1980's. Tyrosinase is involved in the biosynthetic pathway of melanin pigments that is associated with cutaneous pigment disorders essentially linked to (*i*) the hypopigmentation of the skin such as in albinism and vitiligo or (*ii*) the hyperpigmentation of the skin such as in melanoma, melasma and solar lentigo. In addition, tyrosinase is overexpressed during tumorigenesis and has been demonstrated to be a sensitive marker for melanoma. Since the reduction/suppression of melanogenesis is supposed to restore the sensitivity of cancer cells to immuno-, radio- or chemotherapies, the use of selective inhibitors of tyrosinase and related proteins as adjuvants represents a realistic strategy for melanoma therapy. This aspect of the tyrosinase function is addressed by Catherine Belle, Marius Réglier *et al.* in Chapter 2 entitled: "*Transition State Analogue Molecules as Mechanistic Tools and Inhibitors for Tyrosinase.*"

The understanding of tyrosinase functioning is intimately linked to the chemistry of biomimetic models. Indeed, it is thanks to the pioneering work of Kitajima,[1] Solomon,[2] Karlin[3] and Tolman[4] on the dioxygen chemistry of binuclear copper complexes that we precisely understand Tyrosinase's mechanism of action. Two complementary chapters illustrate this aspect of the chemistry of tyrosinase biomimetic models. Chapter 3 by Rabindranath Mukherjee *et al.* is entitled: "Modeling Tyrosinase Activity Using m-Xylyl-based Ligands. Ring Hydroxylation, Reactivity and Theoretical Investigation" and Chapter 4 by Felix Tuczek *et al.* is entitled: "Monooxygenation of phenols by small-molecule models of tyrosinase: Correlations between structure and catalytic activity". Both chapters report on discoveries on bioinspired catalysts that have contributed to a better understanding of tyrosinase mechanism, in particular thanks to reactivity studies, intermediate trapping, spectroscopic methods, and theoretical calculations. Achievements in the development of catalytic bioinspired systems for phenol or catechol oxidation are also described.

The past two decades have seen the emergence of two new copper-containing monooxygenases having a peculiar interest in sustainable chemistry, *i.e.* the mysterious particulate methane mono-oxygenase (pMMO)[5] and the Lytic Polysaccharide monooxygenase (LPMO). While pMMO which converts methane into methanol is far from having revealed all its secrets, LPMOs are well characterized mononuclear copper-containing enzyme including their uses in biotechnology for the development of biofuel production and biosourced chemicals. In an open debate on LPMO active copper/oxygen species, the synthesis of small copper mimics reproducing LPMO structure and activity is in full development and the subject of several publications in the literature. Chapter 5 by Ivan Castillo entitled "Inorganic Models of Lytic Polysaccharide Monooxygenases" addresses this topic.

Studies of copper enzymes mechanism are closely linked to the identification of reactive species.[6] The development of spectroscopic methods coupled with electrochemistry, including at low temperature, is of high importance for the study of the redox properties of transient copper-oxygen species relevant for intermediates in biological processes. These techniques allow accessing important thermodynamical parameters. Chapter 6 from Nicolas Le Poul on *"Electrochemistry and spectroelectrochemistry of copper-oxygen adducts"* illustrates recent achievements in the field.

Finally, inspired by copper enzymes, chemists have developed peptide mimics to bind metal ion while adopting more complex three-dimensional folds. In particular, peptoids — N-substituted glycine oligomers- are bioinspired biopolymers capable of folding into well-defined three-dimensional structures in solution. Chapter 7 entitled "Structure and Function of Cu-Peptoid Complexes" and written by Galia Maayan *et al.* describes the emergence of peptoids as multifaceted ligands for copper and the use of such peptoid-Cu complexes for bioinspired catalysis.

This volume aims to provide interested readers with a selected summary of advances in bioinorganic copper chemistry. We hope that the selection of topics will inspire researchers who wish to undertake research in the bioinorganic field of copper.

A Jalila Simaan
Marius Réglier
July 2023

References

1. Kitajima, N., Fujisawa, K., Morooka, Y., Toriumi, K. μ-η^2:η^2-Peroxo binuclear copper complex, [Cu(HB(3,5-(Me$_2$CH)$_2$pz)$_3$)]$_2$(O$_2$). *J. Am. Chem. Soc.* **111**, 8975–8976 (1989).

2. Solomon, E. I., Sundaram, U. M., Machonkin, T. E. Multicopper Oxidases and Oxygenases. *Chem. Rev.* **96**, 2563–2606 (1996).

3. Karlin, K. D., Kaderli, S., Zuberbühler, A. D. Kinetics and Thermodynamics of Copper(I)/Dioxygen Interaction. *Acc. Chem. Res.* **30**, 139–147 (1997).

4. Halfen, J. A. *et al.* Reversible cleavage and formation of the dioxygen O-O bond within a dicopper complex. *Science* **271**, 1397–1400 (1996).

5. Koo, C. W., Tucci, F. J., He, Y., Rosenzweig, A. C. Recovery of particulate methane monooxygenase structure and activity in a lipid bilayer. *Science* **375**, 1287–1291 (2022).

6. Mirica, L. M., Ottenwaelder, X., Stack, T. D. P. Structure and spectroscopy of copper-dioxygen complexes. *Chem. Rev.* **104**, 1013–1045 (2004).

Contents

2 Transition State Analogue Molecules as Mechanistic Tools and Inhibitors for Tyrosinase   **45**

*Clarisse Faure, Amaury du Moulinet d'Hardemare,
Hélène Jamet, Catherine Belle, Elisabetta Bergantino,
Luigi Bubacco, Maurizio Benfatto, A. Jalila Simaan,
and Marius Réglier*

       Alexander Koch, Tobias A. Engesser, Ramona Jurgeleit,
       and Felix Tuczek

Nicolas Le Poul

Ivan Castillo

7 Structure and Function of Cu–Peptoid Complexes   211

Anastasia E. Behar, Pritam Ghosh, and Galia Maayan,

1 Ligands as a Tool to Tune the Toxicity of Cu on Bacteria: from Boosting to Silencing

Lisa Zuily,* Nora Lahrach,* Enrico Falcone,[†]
Merwan Bouraguba,[†] Vincent Lebrun,[†] Elisabeth Lojou,*
Marie-Thérèse Giudici-Orticoni,* Peter Faller,[†,‡] and
Marianne Ilbert*,[‡]

*Aix-Marseille University, CNRS, BIP, UMR 7281, 31 Chemin
Aiguier, 13009 Marseille, France
[†]Biometals and Biological Chemistry, Institut de Chimie (CNRS
UMR7177), Université de Strasbourg, 4 rue B. Pascal, 67081
Strasbourg, France
[‡]pfaller@unistra.fr
[‡]milbert@imm.cnrs.fr

Abstract

Copper is an essential micronutrient for most living beings including
bacteria. This is mainly due to essential roles of catalytic Cu-centers in
enzymes. However, an excess of Cu is toxic and thereby used over centuries as a powerful antimicrobial agent. Bacteria developed several
systems to detect Cu and protect themselves from high intracellular
Cu concentration, namely *via* Cu-extrusion and/or Cu-storage. Several
Cu-detoxification systems have been described and new ones are still

being discovered. Progress has also been made in understanding the molecular mechanism of Cu toxicity by identifying target macromolecules. Nevertheless, the importance of each mechanism is bacteria- and environment-dependent paving the way for new findings.

The chemical reactivity of Cu can be modulated by its coordination environment in a complex with a ligand (L) and this opens large opportunities to tune the Cu-ligand complexes (Cu-L) *via* ligand design. For instance, boosting the toxicity of Cu ions toward bacteria was promoted by several small organic ligands. These ligands can make Cu-L complexes with different properties in terms of coordinating atoms (S, N, and O), Cu to L stoichiometry, ligand denticity, thermodynamic stability, kinetic inertia, and redox potential. However, one common feature shared by Cu-L complexes that efficiently boost Cu toxicity is the ability of Cu-L to cross the two bacterial membranes and to release the Cu in the cytosol. If this property is required to improve bacteria killing is not clear yet, but considering the ever-growing resistance of bacteria, it would be interesting in the future to try to boost Cu toxicity *via* unprecedented mechanisms.

I. Introduction

Before the Great Oxygenation Event, copper (Cu) was found as copper sulfide minerals such as chalcocite (Cu_2S) and chalcopyrite ($CuFeS_2$), impairing its bioavailability. In this anoxic environment, earliest anaerobic prokaryotes were not using copper as a metal center for their enzymes to operate.[1] Apparition of oxygen on earth led to the oxidation of copper, increasing its solubility, hence turning Cu into a bioavailable metal.[2,3] Living organisms had to adapt to this drastic modification of their environment and they subsequently modify their metallome. For instance, in current aerobic organisms, key enzymatic reactions like the reduction of dioxygen into water are performed with copper-containing proteins.[4]

However, such benefits came with its part of inconveniency. Indeed, while being a key element, Cu-chemical properties render this metal highly toxic for living organisms. Living cells had to face this paradox of an essential yet toxic element. General strategies have been developed by bacteria to remove any excess of copper. Nonetheless, saturation of these systems induces cell death. Such powerful antimicrobial activity enables beneficial use of Cu in several applications. A promising strategy is the development of synthetic copper complexes (Cu-ligands, referred herein as Cu-L) in which the ligand can drastically modulate the properties of copper ions and their mode of action. In this chapter, we will mainly review research exploring the impact of copper and copper complexes on bacteria, and their mode of action. Cu-binding peptides are also attractive options as antimicrobial agents that we will further discuss. Overall, accumulation of recent studies in this field opens new avenue toward potential use of Cu-based compounds in medicine as antibacterial agents.

II. Copper, a Biocidal Compound

The potential use of Cu-related compounds as antimicrobial agents is not a recent idea. Indeed, four thousand years ago, Egyptians already used Cu to sterilize water. Other historical documents revealed that during antiquity, Cu was also prescribed to treat some diseases. A more detailed overview of historical facts revealing utilization of Cu as biocidal compound can be found in other reports.[5-7]

Nowadays, Cu is still worldly employed for its biocidal effect to prevent microorganisms spread: materials containing copper are used in wound dressing, filters, hygienic medical devices, and many others (Figure 1).[8]

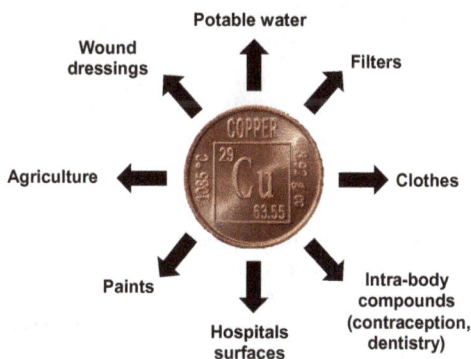

Figure 1. Current applications of copper in our society due to its antimicrobial properties.

As often, the use of the antimicrobial property of copper is not an original human invention, as natural copper-based defense systems were evolved to struggle against invading intracellular organisms. It is now clearly established that increased amount of copper (up to 400 μM) were measured in phagolysosomes after bacteria engulfment.[9–11] In contrast to other metals like iron or manganese that are withhold, the immune system induces production and stimulation of copper pumps (Ctr1 and ATP7A) leading to a copper burst to destroy pathogens (Figure 2).[11,12] Silencing of the gene encoding for ATP7A clearly attenuated bactericidal activity of macrophages.[13] Accordingly, the deletion of *copA* gene (a bacterial gene encoding a pump to export any intracellular copper excess, see section "Copper Homeostasis Systems in Bacteria") renders *Escherichia coli* hypersensitive to macrophage copper-dependent killing.[13]

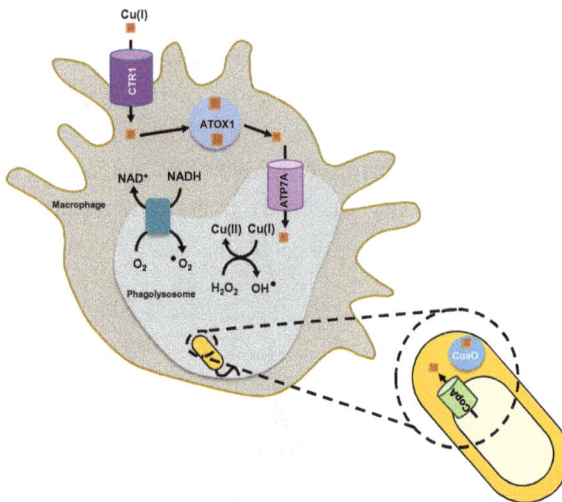

Figure 2. Cu-dependent strategies of the immune system to kill invading microbes. After pathogenic bacteria engulfment, expression of the gene encoding for Ctr1 pump is stimulated, leading to Cu accumulation in the phagolysosome *via* the met-allochaperone Atox1 and the ATPase ATP7A. Pathogenic bacteria (in yellow) will protect themselves from this increase in Cu content by overexpressing specific systems that pump out, transform, or store Cu metals. In this scheme, the CopA pump and the multicopper oxidase CueO are represented as examples.

III. Copper Homeostasis Systems in Bacteria

Copper traffic in cells is tightly controlled because on one hand it is essential for the function of key enzymes but on the other hand it will severely impact cell viability in case of intracellular excess. Cu trafficking occurs mainly in the periplasmic/membrane environment of Gram-negative bacteria where most copper enzymes (superoxide dismutase, cytochrome oxidase, and multicopper oxidase) are localized. How the Cu enters the cells is still unknown as no specific copper transporter has been identified so far. Passive diffusion *via* the outer membrane is suggested but needs to be further confirmed.[14] In some bacteria, the metallophores yersiniabactin, methonobactin or staphylopin are secreted to the extracellular environment where it can bind metals, like copper.[15,16] Such metallophores could play an active role in specific strains to carry metals to the periplasmic compartment.[17] Periplasmic proteins called metallochaperones will then coordinate any periplasmic Cu ions with the dual goal first to carry Cu to specific enzymes that require copper for their function, and second to prevent unwanted Cu reactions with other biomolecules (BMs).[18,19] CusF is, for example, a well-known periplasmic metallochaperone that limits the amount of free Cu and releases it to pumps for Cu export outside the cells.[20]

Any excess of Cu in the environment may lead to an increase of intracellular Cu. To prevent Cu-accumulation and maintain copper homeostasis, specific Cu pumps are overproduced such as CopA or the CusABC system that export Cu out of the different compartments (Figure 3).[18,20] The expression of Cu-specific defense mechanisms is tightly controlled *via* two regulatory systems, one detecting Cu concentration in the periplasmic environment (CusS/CusR two-component systems in *E. coli*, Figure 3), the other detecting an increase in free copper in the cytoplasm (CueR in *E. coli*, Figure 3).[21,22] Other defense mechanisms exist such as the periplasmic enzyme CueO which oxidizes Cu(I) into Cu(II) to prevent the production of toxic reactive oxygen species (ROS) (see section « Cu Reactivity: The Case of ROS Production" for Cu reactivity).[23] Cu-storage proteins have also been discovered. The protein Csp binds up to 80 Cu atoms per tetramer and can be found in the periplasmic space and in the

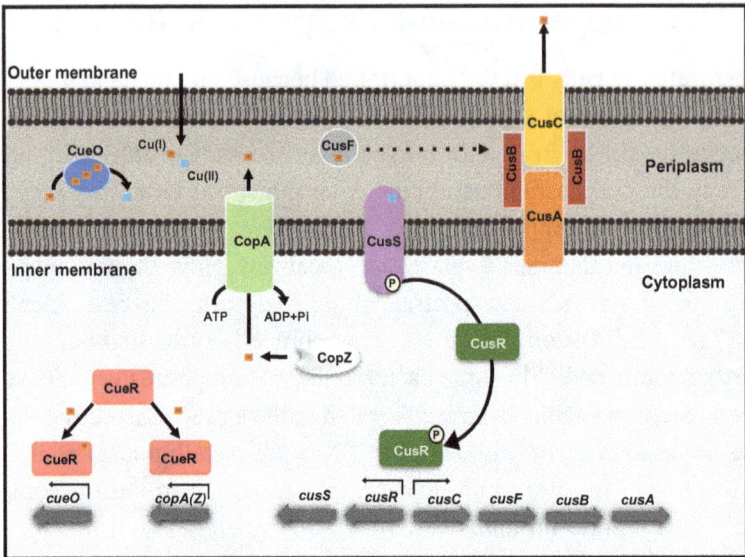

Figure 3. Copper homeostasis systems in *E. coli*. Increase of copper in the periplasmic compartment activates the CusS/CusR two-component systems and leads to the induction of the CusR regulon. CusABC system exports periplasmic copper in the extracellular environment. CusF metallochaperone carries any copper excess to the Cus system to further remove any intracellular copper. Any increase in cytoplasmic intracellular Cu content will activate CueR regulator leading to the overexpression of copA(Z) encoding for the inner membrane pump CopA and the cytoplasmic metallochaperone CopZ which bind cytoplasmic copper and exports it to the periplasmic compartment *via* CopA. CueR also induces the expression of *cueO* gene encoding for a periplasmic enzyme, CueO, which catalyzes the oxidation of Cu(I) into Cu(II).

cytoplasmic compartment of some bacteria.[24,25] In addition, non-protein molecules play a central role to maintain copper homeostasis. The cytoplasmic tripeptide glutathione (GSH) not only maintains a reducing environment but also interacts with copper in excess and confers additional copper tolerance.[26] All these systems found in *E. coli* (Figure 3) are a non-exhaustive list of proteins known to be involved in copper homeostasis in bacteria. Other specialized systems are indeed found in other organisms. For example, in *Rubrivivax gelatinosus*, *copI* is a gene encoding a protein belonging to the cupredoxin family, usually known to be involved in electron transfer.[27] However, in this case, *copI* deletion leads to a strong sensitivity of the strain to copper.[28] The protein CopG recently found in several organisms including *E. coli* and *Pseudomonas aeruginosa* is involved in the cellular protection toward Cu stress.[29] Even in some strains of *E. coli,* a genomic island encoding for the *pco* systems (with an additional multicopper oxidase (PcoA), other periplasmic chaperones (PcoC and PcoE) and a two-component copper sensor (PcoSR)) significantly increases the bacterial resistance to copper.[18,30]

IV. Copper Toxicity: A Phenomenon Dependent on the Bioavailability of Copper and on Bacteria

Copper can be encountered in different habitats as a trace element. However, excess of this metal can be found in contaminated environments after mining process or extensive agriculture treatments. In such cases, environmental bacteria are directly impacted.[31] Even though high level of copper can be measured in such environment, its toxicity will be related to its bioavailability, which is governed by kinetics and thermodynamics. Cu can bind to substances present in a medium or in the soil. Such complexation of the metal renders difficult to estimate the toxicity of an environment based on metal quantification. In soil, for example, the total amount of Cu measured by conventional methods such as inductively coupled plasma–mass spectrometry (ICP-MS) or inductively coupled plasma–optical spectroscopy (ICP-OES) will not provide clear information concerning the amount of bioavailable copper in the sample tested.

Pathogenic bacteria may also encounter high concentrations of copper upon infection of a host. Recent studies on the *Bordetella pertussis* strain have shown that specific copper homeostasis systems have been conserved to withstand copper excess and peroxide stress found in the phagolysosomes.[32] Depending on the complexation properties of the available compounds in its environment, several Cu-derived species may form, with different antimicrobial activities. This may be observed in a simple example: to kill *E. coli* cells, mM concentrations of $CuCl_2$ or $CuSO_4$ are required in Luria Broth medium while only μM are sufficient in a minimal medium (mainly composed of phosphate buffer). Therefore, the most toxic environments correspond to the highest concentration of bioavailable copper which may be quite different from the highest amount of total amount of copper.

In addition, different bacteria species do not survive the same concentration of bioavailable copper. Increasing number of experimental evidences highlight distinct strategies explaining such results. As an example, acido-philic organisms used in bioleaching process can resist copper concentra-tion higher than 100 mM $CuSO_4$;[33,34] in comparison, *E. coli* survives 3–5 mM of $CuSO_4$ in aerobic conditions and in Luria Broth medium.[35] The origin of such resistance has been explained by the duplication of known Cu-resistance genes, new copper-chaperones, surface proteins acting as repellent, and an abundant reserve of inorganic polyphosphate (polyP).[34,36]

V. Specificity of Copper Chemistry

Even though most of the essential d-block metal ions are in the first row (4th period), Cu differs from other metal ions. According to Irving–Williams series, complexes of Cu(II) have the highest stability compared to the other divalent metal ions. Another characteristic is the occurrence of monovalent state Cu(I) (much less common for other biological essen-tial metals like Fe, Mn, *etc.*), which implies a very high thiophilicity. In biological systems, Cu ions are mainly found as Cu(I) (Cu^+, or reduced cuprous state) or Cu(II) (Cu^{2+} or oxidized cupric state). Here, we will rapidly introduce Cu coordination in pure water, coordination geometries found in cells at neutral pH, and then Cu reactivity in aerobic conditions.

A. Cu Coordination Chemistry

Cu(I) and Cu(II) are cations soluble in polar solvents. In water, Cu(II) is the main form encountered, and is coordinated with six water molecules. At higher pH, copper hydroxide $Cu(OH)_2$ is formed, with a very low solu-bility in water. Cu(I) is not stable in pure water: either it is readily oxi-dized to Cu(II) by dioxygen in aerobic conditions, or it disproportionates into Cu(II) and Cu(0) under anaerobic conditions. Hence, to exist as Cu(I) in aqueous solvent, it needs to be stabilized by appropriate ligands. Interestingly, Cu(I) and Cu(II) prefer different coordination geometries, as referred in Table 1, which means that each of these two forms can be sta-bilized by different sets of ligands.[37,38] Worthy of note, Cu(I) is a soft cation (the softest of the essential metal ions) and hence is very thiophilic. In addition, due to the difference in coordination chemistry of Cu(II) and Cu(I), the redox potential can be tuned over a wide range.[39,40]

Table 1. Key coordination properties of Cu(I) and Cu(II) to understand their behavior in biological environment.

	Cu(I)	Cu(II)
Preferred geometry	Digonal, trigonal, tetrahedral, or penta-coordinated	Square planar, with one or two weaker axial ligands
Preferred ligands	Thiolates > thioether, amines > oxygen	Amines, sulfur > oxygen
Stability in aerobic condition	Easily oxidized	Stable
Stability in water (pH 7)	Not stable *disproportionation*	Stable
Solubility in water	$K_s^{CuOH} = 1.2\ 10^{-6(a)}$	$(K_s^{Cu(OH)_2} = 2.2 \times 10^{-20})^b$
Interaction with thiols	Can bind strongly to thiols	Is reduced to Cu(I) by thiols

$^a\ K_s^{CuOH} = \dfrac{[Cu^+] \times [OH^-]}{[CuOH]}$

$^b\ K_s^{Cu(OH)_2} = \dfrac{[Cu^{2+}] \times [OH^-]^2}{[Cu(OH)_2]}$

B. Cu Reactivity: The Case of ROS Production

Cu, like Fe, is a metal ion catalyzing the Fenton reaction. Cu(I) will catalyze the formation of highly toxic hydroxyl radicals (HO$^{\bullet}$) from hydrogen peroxide, a by-product of the respiratory chain. The Cu(II) generated can be reduced by reducing agent (*e.g.*, ascorbate, glutathione (GSH) or any other thiol found in cells). Cu(I) will then further react with either hydrogen peroxide or directly O$_2$, as described in the following series of reactions ("red" denotes a reducing agent and "ox" its oxidized form):

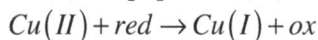

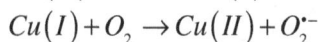

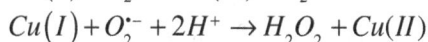

$$Cu(I) + H_2O_2 \rightarrow Cu(II) + HO^{\bullet} + HO^-$$
$$Cu(II) + red \rightarrow Cu(I) + ox$$
$$Cu(I) + O_2 \rightarrow Cu(II) + O_2^{\bullet-}$$
$$Cu(I) + O_2^{\bullet-} + 2H^+ \rightarrow H_2O_2 + Cu(II)$$

Non-controlled Cu amount can thus be a source of reactive oxygen species (ROS).[41] To prevent such reaction to occur, cells maintain at a low level both Cu(I) and Cu(II) species.

C. *In Vivo* Complexation of Cu

In cell, both Cu(I) and Cu(II) are present, in tight interaction with BMs (mainly proteins), which can either stabilize Cu(I) or Cu(II) or cycle Cu(I)/Cu(II) in redox active enzymes. Cu(II) and Cu(I) are found in the periplasmic environment of Gram-negative bacteria, whereas Cu(I) predominates in the cytosol. The presence of thiols (as GSH) in mM concentration explains why copper can be maintained in its reduced form.[42] Under growth condition with low level of Cu in the media, intracellular free Cu(I) is estimated to be extremely low. Around 10^{-21} M Cu(I) is found in the bacteria cytoplasm, compared to 10^{-15} M Zn.[43,44] Actually, keeping intracellular free copper concentration as low as possible constitutes one of the main strategies to prevent intracellular Cu toxicity.

VI. Mechanisms of Cu Toxicity: Macromolecules Targeted by Copper

The bactericide property of copper is well-established. However, how exactly copper leads to cell death still requires intensive investigation. A pleiotropic impact of copper emerged as the most likely scenario (Figure 4). *In vivo* and *in vitro* studies gathered several evidences that Cu acts on any type of biological macromolecules. As previously mentioned, in aerobic conditions Cu is known to produce *via* the Fenton reaction highly toxic hydroxyl radicals, well known to lead to severe damages on proteins, but also on DNA and lipids. For long time, Cu toxicity was believed to be mainly mediated by ROS production. However, it became difficult to explain bacteria sensitivity toward Cu under anaerobic conditions. Remarkably, Imlay *et al.* were able to demonstrate in *E. coli* that excess of Cu under anaerobic conditions lead to mismetallation of key enzymes.[45] Fe–S clusters are particularly vulnerable to Cu-transmetallation, in agreement with the high thiophilicity of Cu(I). Nonetheless, it starts to be accepted that accumulation of several intracellular damages (see Figure 4, and explained below) explains the high level of Cu toxicity toward all cell's kind.

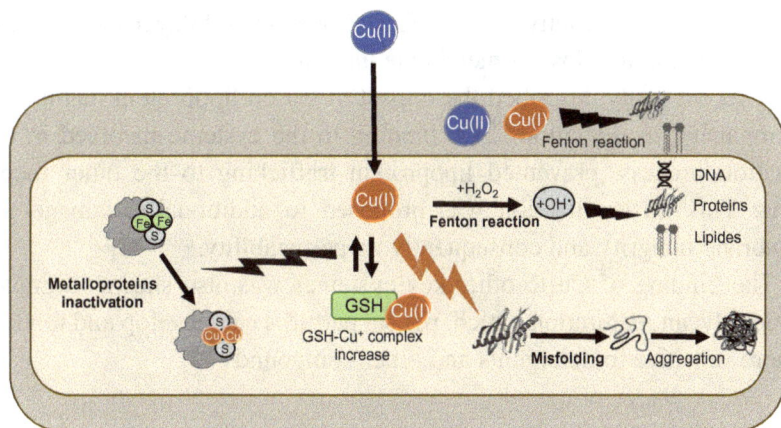

Figure 4. Intracellular Cu impacts. Copper enters the bacterial cell *via* an unknown pathway. In aerobic conditions, copper can be found in the periplasmic compartment as Cu(I) or Cu(II). However, in the reducing environment of the cytoplasm, Cu(II) will be rapidly reduced into Cu(I). Cu(I) in both compartments can be involved in Fenton-type reactions to produce hydroxyl radicals that are highly reactive and toxic to macromolecules. Copper can also have a direct impact on lipids, proteins, and nucleic acid by altering their structure. Free copper or probably the accumulation of glutathione-copper complexes will interfere with metalloproteins by the displacement of the physiological metal and the perturbation of the metallocenter assembly.

A. Membranes

Cellular membrane is, in fact, the first component Cu encounters, and lipids have been described as one of the Cu targets. Nevertheless, the action of Cu on membranes appears to be related to indirect effect *via* ROS production or *via* cysteine-binding to proteins, hence, leading to the modification of proteins and/or lipids that will end in the destabilization of the membrane integrity.

Membrane phospholipids such as phosphatidyl ethanolamine (PE) or phosphatidyl serine (PS) have been shown to coordinate Cu(II) ions *via* reactive groups such as the primary amine of PE found at its head group.[46,47] Hence, the production of ROS is locally increased explaining the high level of lipid oxidation in presence of Cu(II).[47] Other reports confirm that Cu ions bound to the membrane will induce the production of hydroxyl radicals, which will react with lipids and induce lipid peroxidation of the double bonds of unsaturated fatty acids.[48] This phospholipid bilayer modification will further lead to a loss of membrane integrity.

A recent study described the impact of Cu on lipoprotein maturation by impairing its acylation.[49] Cu binding to the cysteine involved in the acylation process, prevented lipoprotein trafficking to the outer membrane. This mislocalization was proposed to additionally damage the membrane integrity and consequently its permeability.

The binding of Cu to other key cysteines was also shown to impair peptidoglycan maturation which, per se, perturbs cell envelop and renders it more sensitive to detergents and other compounds.[50]

B. DNA

Cu toxicity has long been attributed to DNA damages. Indeed, several experimental evidences demonstrate *in vitro* a clear denaturation of DNA fragments in presence of increasing amount of Cu(II).[51] By binding to phosphate and nitrogen ligands, Cu prevents the correct folding of DNA.[52,53] Once bound, Cu further reacts and induces DNA break *via* the production of ROS.[54] However, all these earlier studies were performed in a non-cellular environment. In an environment containing millimolar GSH concentrations, the affinity of DNA to the relevant redox state Cu(I) is low compared to Cu(I) binding to GSH and to other thiolate or sulfide containing proteins (such as those containing FeS clusters). The stronger complex DNA-Cu(II) would never form *in vivo* as Cu(II) will be reduced immediately by GSH. In addition, in *E. coli*, most of Cu ions are found in the periplasm. Very little amount may enter in the cytoplasm where the formation of hydroxyl radicals is unlikely given the high concentration of GSH in this compartment and its potency to stabilize Cu(I). In accordance, Macomber *et al.* demonstrated the absence of severe DNA damages in *E. coli* after Cu treatment.[35] In fact, the Cu binding capacity of DNA might be used by bacteria to scavenge extracellular Cu ions and prevent Cu intracellular damages. Dalecki *et al.*[55] proposed that such strategy could be used by bacteria pathogens to protect themselves against copper by the release of extracellular DNA once inside eukaryotic cells[56] or once they formed biofilm matrix notably composed of extracellular DNA.[57] Further studies might be of interest to better define the role played by DNA among bacterial strategies to survive copper excess.

C. Proteins

Proteins are one of the most sensitive targets to Cu overload in bacteria cells. Cu ions bind to O-, N-, and S- donors, elements that are found in the protein backbone as well as in several amino acid side chains (such as cysteine, methionine, histidine, *etc.*). Such variability of interactions leads to a multitude of impact on proteins and, *in fine*, on cells.

In aerobic conditions, intracellular Cu excess leads to severe oxidative damage linked to hydroxyl radical production *via* the Fenton reaction, as well as to the oxidation of cysteine residues that may form unwanted disulfide bonds. Deletion of periplasmic repair system known to fix abnormal disulfide bonds, decreases drastically *E. coli* survival upon Cu stress in aerobic conditions.[58] This was not observed under anaerobic conditions, underlying the different mechanism of action of Cu under anaerobic *versus* aerobic conditions.[59] Cu might also inhibit the reduction of disulfide bonds in the periplasm, a process required to mature cytochrome *c* and this has been shown to perturb key pathways impairing cell survival.[28] Recently, methionine oxidation of the periplasmic protein CusF has been reported.[60] These results confirm the production of ROS in the periplasmic environment of *E. coli*. Under these conditions, periplasmic methionine sulfoxide reductase repairs the oxidized methionines of CusF, demonstrating the importance for cells to keep their copper homeostasis systems active by repairing the oxidation of their residues.

In addition, Cu(II) induces protein misfolding *in vitro* and this was recently demonstrated *in vivo*.[59,61,62] As described earlier, Cu(II) metal ion possesses a strong affinity for proteins compared to other metals, reflecting its place in the Irving–Williams series. Furthermore, Cu(I) shows a high affinity for thiols, such as cysteines. Hence, any intracellular concentration increase of Cu could lead to a strong perturbation of Zn-, Fe-, or other metalloproteins, as intracellular concentration of metals have been shown to dictate some proteins metalation.[63,64] Under anaerobic conditions, direct inactivation of metalloproteins by Cu has been demonstrated *in vivo* by Imlay *et al.*[45] The authors clearly showed that the iron–sulfur enzyme (isopropylmalate dehydratase) is one of the first Cu-target in *E. coli*. Cu-induced protein inactivation leads to cell death unless amino acids were added to the media. Inactivation of iron–sulfur enzymes or other metalloproteins, as well as perturbation in iron–sulfur biosynthesis or heme biosynthesis, has been further confirmed by other authors in other microorganisms.[65–68]

Intriguingly, under anaerobic conditions, a lower concentration of Cu in the medium is required to kill *E. coli* cells compared to aerobic treatment.[35] This was at first unexpected as Cu toxicity was thought to act mainly *via* ROS production. Instead, this result further accentuates the adverse impact of intracellular Cu on proteins in a ROS-independent manner. Recently, a correlation between a higher intracellular copper content observed under anaerobic conditions and an increased level of intracellular aggregated-proteins has been clearly demonstrated.[59] This result emphasizes the direct impact of copper on protein folding. Molecular chaperones have been shown to maintain the cellular proteostasis under copper stress conditions highlighting a central role of this family of proteins to protect cell against copper toxicity.[59]

To summarize, the proposed model to explain bacteria death by Cu is more complexed than initially thought. Under aerobic conditions, intracellular copper excess as well as ROS-generated by Cu will interfere with all macromolecules and destabilize several pathways. Accordingly, a recent study highlighted several genes involved in the resistance of *E. coli* toward a prolonged copper exposure confirming the multiple impact of copper on the cell.[69] Under anaerobic conditions, Cu will destroy cells by inducing protein misfolding and/or by mismetalating catalytic sites and this will also lead to cell death. As mentioned in a recent paper by O'Hern *et al.*: "a fundamental understanding of Cu speciation and availability in cells is essential to uncover the cellular consequences of Cu cytotoxicity".[70]

Overall, Cu toxicity in bacterial cells is likely to be multifactorial: ROS production, mismetalation, and binding to key amino acids might be the main features that lead to global cell defect in bacteria. However, in the abovementioned experimental studies, high Cu salt concentration ($CuCl_2$ or $CuSO_4$) was required to observe cell death. With the objective of using Cu as medical treatment, Cu-cytotoxicity needs to be improved. With such aim, Cu-complexes appear as appealing compounds and we will next develop the advances made in this domain.

VII. Chemistry of Copper Complexes

The emergence of antibiotic-resistant bacteria has urged scientists to find alternative killing strategies in replacement to classic compounds.

Developing new antibiotic agents is challenging, time consuming, and might end with the apparition of new resistance mechanisms. Cu is currently reconsidered as therapeutic agent in complex or in combination with antibiotics.[55]

A variety of ligands (hereafter designated as L) can form complexes with Cu ions through the formation of coordination bonds. Inducing drastic changes in the properties of Cu, a ligand can be rationally designed to obtain complexes for specific applications. For example, depending on its chemical groups, the global charge of Cu-L complexes can be negative, positive, or neutral, hence changing the solubility in hydrophilic or hydrophobic environments as well as its affinity toward intracellular components. In this chapter, we will give a rapid overview of the chemical properties of Cu-L that can be tuned by modifying the ligand.

A. Reactivity of Cu Complexes

Since Cu-cytotoxicity can arise from different chemical mechanisms (see Section 1.6), ligands (L) are a good tool to tune it and/or control its specificity. Indeed, ligands will promote or prevent some of the mechanisms listed below.

(a) Redox-based: ligands modulate Cu redox potential (see section "Redox Potential") modifying Cu redox reactivity. Moreover, a redox reaction often involves the binding of the substrate, replacing a labile ligand (*e.g.*, H_2O), prior to the electron transfer. In that case, which is considered common for Cu-L complexes, the ligand (L) might influence the interaction with the substrate, from hindering it, to attracting or directing/orientating a potential substrate.

(b) Catalytic redox-independent: Cu(II) (but not Cu(I)) being a quite strong Lewis acid, biological relevant reactivity could include non-redox reactivity, such as hydrolytic cleavage of BMs *via* the modulation of the pK_a of Cu-bound water. In this case, for instance, a ligand can further change the pK_a. Of note, non-redox catalysis cannot be excluded as a potential mechanism involved in Cu toxicity, but is often considered less likely.

(c) Pure coordination: the type and structure of the ligand (L) in Cu-L can influence how Cu binds to a BM, often a protein (Figure 5). This is

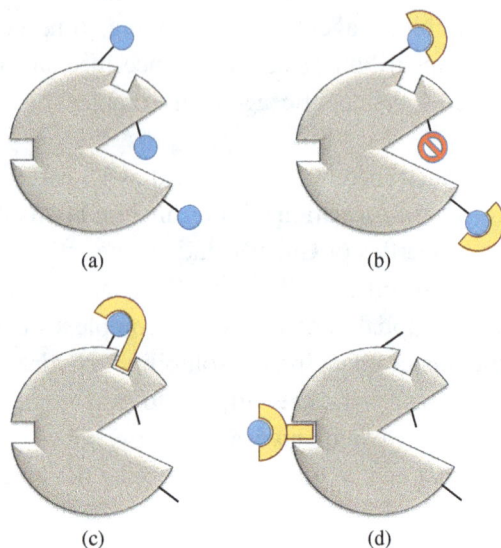

Figure 5. Influence of ligand L on the interaction of Cu with biomolecule: (a) ionic Cu (blue balls) can bind to several coordination sites. L (in yellow) can confer selectivity to Cu binding *via* ternary complex by different means; (b) by steric hindrance of the L that confers selectivity to certain Cu-binding sites, *e.g.*, Cu-L cannot bind to buried sites in a protein; (c) by the interaction of the ligand with the target protein together with a Cu-coordination site of the protein; or (d) by binding of the Cu-L *via* weak interactions (*e.g.*, H-bonds) but without coordination bond between Cu and target protein.

especially relevant for ternary complexes L-Cu-BM. A ligand can modulate the interaction with BM *via* steric hindrance or ligand–protein interactions (Figure 5). As a result, some BMs will be preferred targets based on the nature (*i.e.*, chemical function) and exposition (*i.e.*, accessibility) of their coordinating groups.

B. Stability of the Cu-L Complex

The stability of metal-complexes against dissociation is a key parameter to define their intracellular behavior. Complex stability is often expressed as apparent dissociation constants (K_d), which are measured under a certain condition (pH, buffer, concentration, *etc.*). To compare the affinity among different complexes, these apparent K_d have to be measured at the same pH (often being pH 7.4) without any competitor (such as buffer).

Besides, one must care of the stoichiometry of the complexes before comparing their K_d. When appreciating apparent dissociation constant of complexes, important concepts have to be considered, such as chelate effect, Pearson's principle (hard and soft acids and bases [HSAB]), coordination geometry, and charges (including partial and electronic densities). Of course, another important concern is the physiological context in which the metal-complex will be used. All of this needs to be taken into account in *in vitro* measurements to get a better idea of Cu-L complex intracellular stability and consequently intracellular behavior.

C. Kinetics of the Cu-L Complex

The inertia of the Cu-L complex, *i.e.*, how fast L can dissociate from a Cu-L complex and/or how fast an external ligand L_E can bind to Cu-L, can also play a role. Cu(II) and Cu(I) are quite labile metal ions. They can be exchanged quite rapidly in Cu-L, unless the affinity is very high or the ligands are very bulky and shield Cu. The latter is common in some proteins, but more difficult to obtain with smaller ligands. Here again, the design of the ligands will allow to tune the kinetics of dissociation for the purpose expected.

D. Redox Potential

Cu ions in a Cu-L complex can have very different redox potentials. The main biological redox couple is Cu(I)/Cu(II). Depending on the properties of the ligand (geometry, number, and type of coordination atoms (O, N, S, *etc.*) and charge or electron density of the ligand (see Table 3), the redox potential can vary from below −0.4 up to +0.8 V *versus* NHE. This means that a range of potentials can be reached in which Cu can switch from quasi-inert behavior in biological environment to a high redox activity.

VIII. Impact of Ligands on the Biological Activity of Copper

As previously mentioned, modification of the ligands may change significantly the intrinsic properties of Cu. Drastic intracellular changes and impact on cell survival are expected. Four main strategies drive the development of new ligands toward the tuning and control of the killing efficiency of copper-based drugs: (i) improving membrane permeability, (ii) improving the specificity of the reaction, (iii) tuning the compartment targeted, or (iv) tuning the mechanism of action. In this chapter, we will discuss the different intracellular routes Cu-complexes could undertake.

A. Localization

Cu-L cellular localization is largely influenced by its charge: (i) hydrophobic Cu-L might localize into the membranes, (ii) less hydrophobic (yet not too hydrophilic) ones could cross the membranes, and (iii) highly charged ones are expected not to passively penetrate membranes. Noteworthy, a hydrophobic complex (neutral global charge, hydrophobic ligand) might have low solubility in aqueous media. Therefore, experimenters must care about precipitation.

B. Presence in Biology of Potentially Competing Ligands for Cu(II)

As mentioned earlier, biological media possess high amount of potential metal ligands such as glutathione, proteins, *etc.* Concerning the extracellular and the periplasmic environments, where Cu(II) is stable, weak Cu-L can dissociate rapidly. In such case, any biological effects will be due to Cu and/or ligand only. To maintain intracellular Cu-L complex, a strong affinity is required to prevent dissociation. For instance, if a Cu-L is conceived to target bacteria in humans *via* the blood, the Cu-L has to resist to Cu-binding proteins present in that fluid, such as serum albumin (Cu(II) $K_d = 10^{-13}$ M).[71]

As a general rule of thumb, monodentate and bidentate ligands are generally too weak and will be dissociated easily, unless a stable ternary complex with an external ligand is formed (see section "Reactivity of the Free Ligand in Bacteria", Table 2). To keep the integrity of the initial Cu-L complex, stronger ligands, tridentate or tetradentate, should be favored. It is noteworthy that ligands coordinating metals *via* only oxygen atoms have generally a moderate affinity for Cu(II).

While in the periplasm Cu(II)-L remains oxidized, Cu(II)-L entering the cytosol will be reduced in this compartment, which contains high amount of thiols, *e.g.*, 1–10 mM of reduced GSH. Thiols are very efficient reducing agents for Cu(II) and Cu(II)-L and quite strong ligands for Cu(I) (apparent K_d of low μM Cu(I) to 1–10 mM GSH is ~10^{-16} M[39]). Actually, most of the Cu-L used as antimicrobial agents are relatively strong Cu(II) ligands, stable in the periplasmic compartment. Once such complex enters the cytosol, Cu-L will be reduced and dissociated if L is a weak ligand for Cu(I). As the coordination chemistry of Cu(II) and Cu(I) are very different, a ligand that strongly binds Cu(II) can be a weak ligand for Cu(I). Ligands that enable such transport of Cu into the cytosol have been referred to as ionophores (Figure 6). Here, the toxicity of the Cu-L will be in fact due to the Cu(I) and ligand released into the cytoplasm rather than directly to the Cu-L complex.

Highly stable Cu(II)-L complexes do exist. They are even maintained in the cytosol in presence of thiol compounds. However, in such a case, they have no redox activity and are less toxic[72] (*e.g.*, Cu-atsm, see section "Bis-Thiosemicarbazones (atsm, gtsm)").

Furthermore, the formation of ternary complexes can occur already outside the bacteria, in the cell medium or in the blood of the host, for example. This can even more modulate the fate and activity of Cu-L. Indeed, extracellular ligands (*e.g.*, amino acids and proteins) may either act as "scavenger/trap" of Cu-L, weakening its effect, or enhance its cellular uptake and hence its activity.

C. Presence in Biology of Potentially Competing Ligands for Cu(I)

Cu(I)-L complexes are unstable in presence of O_2. They can be rapidly oxidized extracellularly or in the periplasm. In such case, dissociation of the Cu-L complex will occur because strong Cu(I)-ligands are often weaker Cu(II)-ligands. However, O_2-stable Cu(I)-L exists, and if they are hydrophobic, they can passively enter in the periplasm and/or cytosol. They can be stable in the cytosol. The requirement is a stability constant high enough to resist GSH (K_d < fM) or to prevent the activation of the

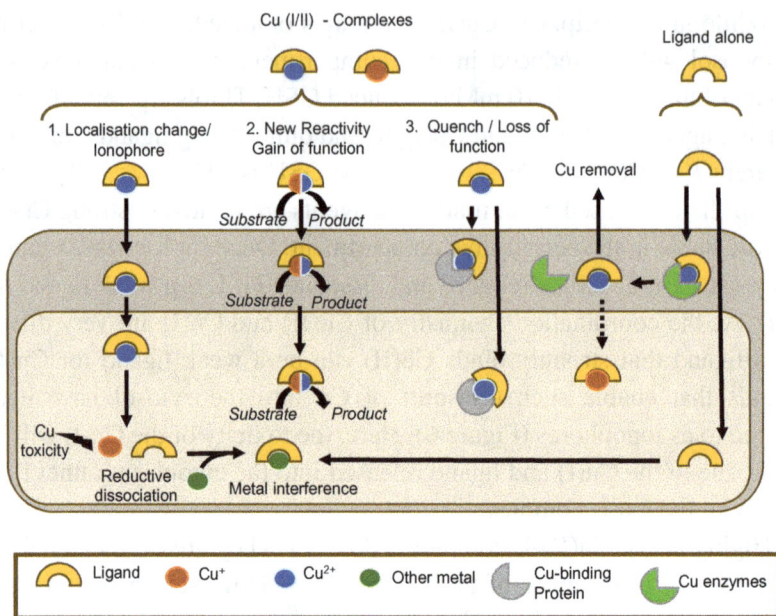

Figure 6. Possible routes and activities of Cu bound to ligands. The left part of the panel describes the activity of a Cu(I/II)-L encountering the bacteria with three conceptual routes. On the right part of the panel, the Cu-free ligand (L) encounters the bacteria, but once inside the cells, it will interact with endogenous Cu enzymes or interfere with intracellular metal homeostasis. For Cu-L, route 1 describes the ionophore activity (also called "mechanism I" in this review). The role of the ligand is to help the Cu to pass the membranes hence increasing the intracellular Cu concentration. Very often the Cu dissociates from the ligand triggered by Cu(II) reduction. The released ligand could bind other metals or interfere with other targets as well the Cu. Route 2 consists of a new function of the Cu-L. Classically, this would be a Cu-L catalysing a reaction, most likely the production of ROS (also called "mechanism II" in this review). In route 3, the Cu-L inhibits an essential function (loss of function) such as binding to an essential cysteine in a cytoplasmic or periplasmic enzyme.

transcription factor CueR (K_d < zM) and the consequent expression of the Cu extrusion machinery. Such complex might not have strong intracellular toxic effect. Otherwise, Cu(I)-L can dissociate in the cytosol, exerting an ionophore activity.

D. Reactivity of Cu-L in Bacteria

Based on several experimental evidences, some Cu-L show a clear increased bacterial toxicity in comparison to free copper (see section

"Copper Complexes as Antimicrobial Agents, a Non-Exhaustive List"). This can be explained by our previous description of the reactivity of intracellular copper complexes, which can be summarized as follows (Figure 6):

(a) Cu-L ionophore activity: the complexation of Cu serves to increase Cu transport efficiency through the membranes. Higher intracellular accumulation of free released Cu can easily explain the higher level of toxicity. In addition, the dissociated ligand could also perturb intracellular pathways. Most of the Cu-L characterized so far, belong to this category (see section "Copper Complexes as Antimicrobial Agents, a Non-Exhaustive List").

(b) Gain of function *via* redox reactions: some Cu-L can undergo redox reactions, like reducing O_2 to form ROS or oxidizing essential BMs (cysteines, GSH, NADH, *etc.*). To have a high redox activity, the redox potential should be between ~0 and 0.3 V *versus* NHE, and the Cu-L has to rapidly cycle between Cu(I) and Cu(II). In this case, there is tradeoff between stability and reactivity. Indeed, to obtain a stable and highly reactive Cu-L complex, the ligand needs to bind strongly both redox states (Cu(I) and Cu(II)). This represents a real challenge as the two oxidation states prefer drastically different coordination spheres (in terms of geometry mainly, but also composition).[72]

(c) Loss of function *via* binding: a Cu-L complex may be stable enough to react/bind, for instance, to an essential residue of an enzyme, such as cysteine (for Cu(I)-L or Cu(II)-L) or histidine (most likely for Cu(II)-L), which will lead to a loss of function. In that case, one of the ligands of the initial complex must be labile to be replaced by the targeted protein residue. It is also possible that Cu-L binds to a target *via* weak bonds (*e.g.*, hydrophobic interaction and hydrogen bonds) in a pocket of its target and inactivates it (see case of Bis-thiosemicarbazones (bTSC) below, section "Bis-Thiosemicarbazones (atsm, gtsm)").

E. Reactivity of the Free Ligand in Bacteria

Instead of providing the Cu-L complex directly to the cell, it could be possible to form it intracellularly by adding the ligands in the media that would bind endogenous Cu in the bacteria (Figure 6). This may have biological activity *via* two possible mechanisms:

(a) Such interaction could perturb intracellular copper homeostasis and destabilize essential metabolic pathways, in addition to the intrinsic activity of Cu-L. An important prerequisite is the strong affinity of the ligand L for Cu. In the cytosol this is very challenging, as intrinsic protein (CueR) has already a very strong affinity (zM range[43]), and GSH binds Cu(I) in the low μM range.[42] In the periplasmic compartment, strong competitors are available. Indeed, even proteins that do not bind Cu under physiological conditions can have quite strong affinity for Cu(II) to compete against L. As an example, proteins containing a His in 2nd or 3rd position (from the first amino acid in *N*-terminal) which are quite common in biology (albumin, histatin, antimicrobial peptides (AMPs), *etc.*) have K_d in the range of 10^{-12} M to $10^{-14.5}$ M.[73] Therefore, the design of L may be quite challenging, as its affinity must be suited to compete against endogenous Cu(II)- or Cu(I)-ligands.

(b) An interesting and underexplored case is the formation of a ternary complex L-Cu-protein. Here, a ligand added to the media could, after cellular insertion, bind to a copper center found in an essential enzyme. This will impair the function of the enzyme and kill the cells. Noteworthy, this strategy converges with previously described one, consisting in inhibiting an enzyme by targeting a cysteine or histidine in its active site with Cu-L. In this strategy, it is important to consider the denticity of the ligand and the favored stoichiometry of Cu/L complex(es) (Table 2). For, Cu(II), a tetradentate L has no vacant equatorial coordination site available for a strong bond, and hence ternary L-Cu-protein is not stable. In contrast, a bidentate ligand to Cu(II) can form two strong equatorial bonds in L-Cu-protein. This could happen also after the release of one L in a Cu(II)-L_2, for example in Cu(II)-1,10-Phen$_2$ (see section "Phenanthrolines").

As just discussed, several parameters can be modified to specifically modulate the properties of a Cu-complex. Even though several Cu-L have already been designed and show interesting antibacterial properties, the next chapter will provide some relevant examples, all the possibilities

Table 2.   Cu(II)-L complexes general properties.

Ligand type	Bidentate	Tridentate	Tetradentate
Complexes formed with Cu(II)			
Thermodynamic stability			High
Kinetics (ligand exchange)	Fast		
Formation of ternary complexes (*e.g.*, with proteins)			
Reduction by physiological reducing agent	Easy/Fast		
Example (for more details see section "Copper Complexes as Antimicrobial Agents, a Non-Exhaustive List")	Cu(II)-1,10-Phen$_2$ Cu(II)-HQ$_2$	Cu-terpy[a]	Cu(II)-atsm Cu(II)-gtsm

In green L$_E$ correspond to external ligand. External ligands are classically water (or other solvents) or anions during the preparation of the complexes (*e.g.*, halides). Upon addition in cell culture medium or application to living organisms, they are supposed to be exchanged easily with small molecules (water, amino acid, and anions) or with proteins and hence form the ternary complex (L-Cu-BM).
a - .74

presented above have not yet been fully exploited and further efforts are required to improve our knowledge on the impact of diverse Cu-L on bacteria.

IX. Copper Complexes as Antimicrobial Agents, a Non-Exhaustive List

Hereafter, we will discuss some of the most prominent class of Cu-complexes (Cu-L) used as antibacterial agent. Some of these complexes are already used to treat cancer or neurogeneration but they appear well-suited to fight microbes as a very low amount is sufficient to kill them.

A. Dithiocarbamates Compounds

Dithiocarbamates (DTC) are a class of metal chelators that can be formed from thiuram disulfides by a two-electron reduction (Figure 7). They bind Cu(II) bidentately and can form Cu(II)-DTC and Cu(II)-DTC$_2$ complexes. Cu(II)-DTC$_2$ is neutral and the redox potential is quite low (Table 3). Thiuram disulfides have a much lower affinity for copper ions than DTCs. Thus, in terms of copper binding, thiuram disulfides are prodrugs, and need to be activated by reduction, which is supposed to occur mainly intracellularly by thiols. The most prominent example is the antabuse drug disulfiram (a neutral thiuram disulfide) that upon reduction yields the active DTC drug, diethyldithiocarbamate (DEDTC).

Thiuram disulfides, or most likely the intracellular compound DTC, have been shown to have antibacterial activity on *Staphylococcus aureus*,[75] on *Mycobacterium tuberculosis*,[76] and on *P. aeruginosa*.[77] A link

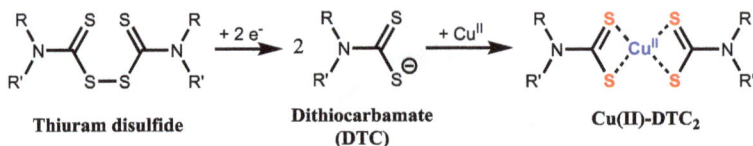

Thiuram disulfide   Dithiocarbamate   Cu(II)-DTC$_2$
                                 (DTC)

Figure 7. Reduction of thiuram disulfides to dithiocarbamates and subsequent Cu(II)-chelation. Several derivatives are studied depending on the R/R'. (R=R'= -CH$_2$-CH$_3$ for DETC, R/R' = cyclic -CH$_2$-CH$_2$-CH$_2$-CH$_2$- for pyrrolidinedithiocarbamate (PDTC)).

between the antimicrobial activity and DTC ability to bind copper was proposed but not clearly demonstrated. As a thiol agent, DTC might react with cysteine from key enzymes to form a thioester.[77] DTC might also be able to bind Cu(II) in the oxidizing periplasmic compartment of Gram-negative bacteria and per se perturbs Cu trafficking toward Cu-enzymes. Last, DTC might transport Cu(II) into the cytoplasm where it is expected to be released as Cu(I) due to the high thiol content.[72] Coherently, in this reducing environment of bacteria cytoplasm, DTC might not form a complex with Cu(I) as it will not compete with GSH found in mM concentration.[72]

Instead of adding only the ligand DTC, direct addition of Cu(II)-DTC$_2$ complex in the media has a strong antimicrobial effect. This activity can be explained by intracellular accumulation of Cu due to ionophoric activity (mechanism I). Accordingly, strong Cu-dependent antimicrobial activity of disulfiram/DTC was reported against *M. tuberculosis*, with a minimum inhibitory concentration (MIC) around 0.3 μM,[78] and against *S. pneumoniae*.[79] The authors further showed the ability of Cu(II)-DTC$_2$ complex to cross bacteria membranes and inactivate cytoplasmic enzymes.[78] In such condition, intracellular Cu stress response is induced. This confirms that Cu(II)-DTC$_2$ entering the cytoplasm might release high Cu(I) concentration which perturbs metabolic pathways. Such Cu-dependent growth inhibition of disulfiram/DTC has also been observed recently on mollicutes.[80] The addition of a Cu(II) chelator in the media completely modifies the MIC values of disulfiram from nanomolar to micromolar range. This result further confirms the critical role of Cu for disulfiram antimicrobial activity.

Besides, Cu(II)-DTC$_2$ has been shown to interact with zinc finger proteins in cancer cells, and to oxidize Zn(II)-thiolates to form disulfide.[81,82] Even though this mechanism has not been reported in bacterial cells, its occurrence cannot be ruled out.

B. 8-Hydroxychinoline (8-HQ) Compounds

8-HQs are neutral quite hydrophobic compounds able to cross membranes passively.[83,84] They have been described as antimicrobial agent and have been studied on several organisms demonstrating their efficiency.[83,85] The 5-chloro-7-iodo-8-HQ, called clioquinol, was developed in 1899 as an antiseptic and later used as an amebicide, but withdrawn from the market in the 70s due to side effects.[86] A high-throughput physiological screen made against *M. tuberculosis* identified 8-HQ compounds as efficient molecules.[87] Although the antimicrobial mechanism is multifaceted, 8-HQ interaction with metal ions including Cu seems to play a role. Addition of Cu ions triggered killing fungal pathogen *Cryptococcus neoformans* by a 8-HQ, and further addition of the strong Cu-chelator BCS forming a membrane-impermeable Cu–BCS$_2$ complex reversed the toxicity, in line with an important contribution of a Cu-ionophore activity of 8-HQ.[83]

8-HQ can form Cu(II)-HQ and Cu(II)-HQ$_2$ complexes (Figure 8). The affinity of Cu(II) for HQ$_2$ is quite high at neutral pH, whereas the metal ions binding *via* a phenolate (hard base) suggests a relative low affinity for Cu(I) (soft acid). Despite the quite low redox potential of Cu(II)-HQ$_2$, it can be readily reduced by the highly concentrated GSH tripeptide found in the cytoplasm, likely *via* the formation of a ternary complex. Cu-complexes formation with 8-HQs will consequently depend on the bacteria compartment, *i.e.*, possible with Cu(II) in the periplasm or in the environment, but not with Cu(I) in the cytoplasm. Consequently, 8-HQ is also well suited to serve as a ionophore (mechanism I). Such a mechanism of action was indeed proposed by Festa *et al.*[83] to explain 8-HQ and Cu-8-HQ toxicity on *C. neoformans*. The authors observed an increase in intracellular bioavailable Cu content which can then affect intracellular pathway either by acting directly on proteins or by generating toxic ROS.

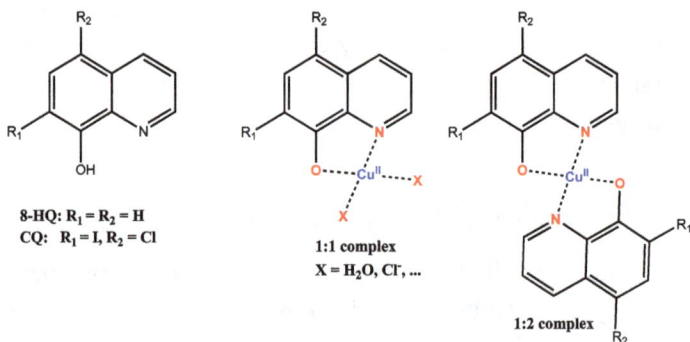

Figure 8. Structure of 8-HQs and their Cu(II)-complexes.

However, 8-HQ toxicity varies depending on the organism tested.[83] Species-specific targets and different Cu resistance strategies may explain this diversity, requiring further investigations.

It is also important to mention that 8-HQ with their hard phenolate ligand are good Fe(III) binders compared to DTC. Dissociation of Cu(II) might induce some cross talk with Fe(III) pathways which could also be a process involved in 8-HQ toxicity.

C. Phenanthrolines

Phenanthrolines (Phen) are bidentate ligands forming both 1:1 and 1:2 complexes with Cu(I) and Cu(II) (see Figure 9). 2,9-unsubstituted

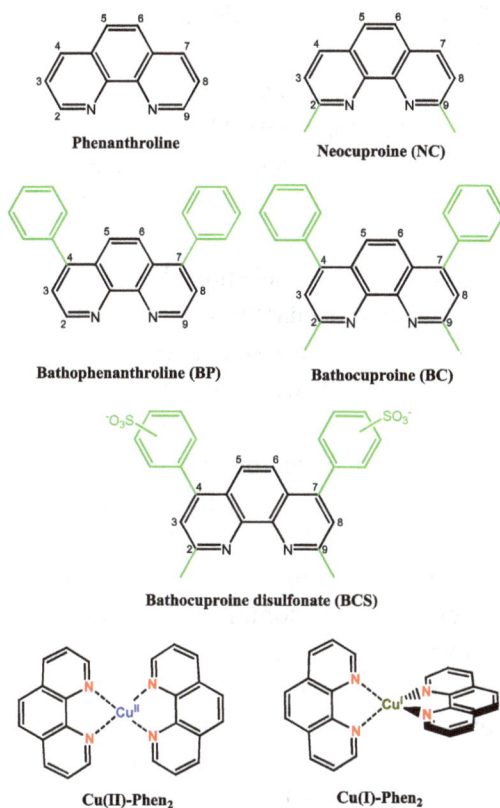

Figure 9. Structure of phenanthroline, its derivatives and their Cu(II)-complexes. Structural difference between the ligands are highlighted in green.

phenanthrolines (*e.g.*, 1,10-phenanthroline, bathophenanthroline) have generally a high affinity for both Cu redox states.

In contrast to 8-HQs and DTCs, these phenanthrolines can undergo rapid redox-cycling between Cu(I) and Cu(II) and are potentially very competent in ROS production (mechanism II). However, *in vivo,* they likely act only as ionophores (mechanism I) rather than pro-oxidant. For instance, the Cu(I) and Cu(II)-complexes with neutral phenanthrolines can passively cross the membranes, despite the positive charge of the complex formed. Then, Cu-(Phen)$_2$ is rapidly reduced by GSH, at least one Phen dissociates and a ternary complex Phen-Cu(I)-GSH might form.[88] However, other proteins with stronger Cu(I) affinity than GSH might be able to totally subtract Cu(I) from Cu-phenanthroline complexes. This has been shown for mammalian metallothioneins,[72] but might be also the case for bacterial metallothioneins or for copper storage proteins such as Csp3.[25]

It is important to distinguish phenanthrolines according to the substitution at positions 2 and 9 (Figure 9). Methyl (or larger) groups at these two positions, such as in bathocuproine (BC), neocuproine (NC), cuproin, or bathocuproine disulfonate (BCS), stabilize Cu(I) over Cu(II) in the Cu-Phen$_2$. In other words, they have a stronger Cu(I)-affinity, but lower Cu(II)-affinity compared to 2,9 unsubstituted phenanthrolines. The bulky methyl-groups force the two phenanthrolines into a tetrahedral coordination of the metal ion, a geometry suited for Cu(I) but not for Cu(II). Related to that, Cu(I)-complexes with two 2,9-substituted phenanthrolines (*e.g.* Cu(I)-BC$_2$) are difficult to oxidize and are generally air-stable.

It is a general feature that the apparent affinity for a metal ion in a 1:2 (metal:ligand) complex has a more drastic dependence on the ligand concentration compared to a 1:1 complex. Hence, 2,9 substituted phenanthrolines that form a 1:2 complex can become stronger than Cu(I)-proteins (that often are 1:1 complexes) at higher concentrations > 0.1 mM.[89] However, at lower ligand concentrations, Cu(I) might dissociate from 1:2 complexes as shown for 10 μM Cu(I)-BCS$_2$ which dissociated at about 50% in 3 mM GSH and totally with mammalian metallothioneins.[72] Therefore, Cu(I) complexes with two neutral 2,9-substituted phenanthrolines (like BC or NC) can also have ionophore activity at lower concentrations.

The antimicrobial activity of some of these compounds has been demonstrated, which is highly dependent on their ability to cross the membrane. NC exhibited efficient *M. tuberculosis* growth inhibition in presence of increasing amount of copper.[90] Such an effect was also observed on *Mycoplasma* and *Ureaplasma* species.[80] In contrast, bathocuproine disulfonate is membrane impermeable and does not show any effect on *M. tuberculosis* growth.

Even if some of these compounds cannot be used as antimicrobial agent, their ability to tightly bind Cu(I), as NC and BC, have made them widely used as chelating agents *in vitro* or *in vivo* for several experimental purposes.

D. Pyrithione

Pyrithione (PT) is a neutral ligand which can bind Cu(II) as a Cu(II)-PT$_2$ complex, which is neutral as well, due to the loss of the two hydroxylamine protons (Figure 10). Thus, in terms of charge, PT has properties suited for ionophore activity (mechanism I), as Cu(II)-PT$_2$ can enter the cytoplasm and PT exit it. The affinity of Cu(I) to PT is unknown, but according to Pearson theory, Cu(I) has a low affinity for N-OH. This means that once reduced in the cytoplasm, Cu(I) will dissociate from PT. This was further confirmed by a recent study,[62] where the authors demonstrated the intracellular increase of copper in *E. coli* after treatment with a mix of Cu and PT and in comparison, with Cu alone. Cu-induced

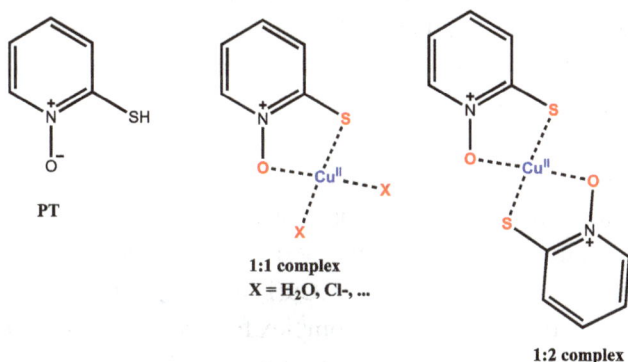

PT

1:1 complex
X = H$_2$O, Cl-, ...

1:2 complex

Figure 10. Structure of PT and its Cu(II)-complexes.

cell death is amplified by Cu-PT$_2$ which most likely might overwhelm copper homeostasis systems. Cu-PT$_2$ treatment was also tested on *Neisseria gonorrhoeae* which is highly sensitive to this complex with a MIC similar to MIC observed for Cu-gtsm compounds (see section "Bis-Thiosemicarbazones (atsm, gtsm)).[91] On *Klebsiella pneumoniae*, the addition of PT with either Zn or Cu increases the antibiotic activity of amikacin. Here, the authors propose an ionophore activity of this compound.[92]

Of note, the structure of Cu-PT is reminiscent of the natural copper-containing antimicrobial agent Fluopsin C, in which Cu is coordinated by two units of N-methylthiohydroxamate. Interestingly, both Cu-PT and Fluopsin C show broad antimicrobial activity against Gram-negative and Gram-positive bacteria, as well as multidrug-resistant strains. By contrast, they are not effective against *P. aeruginosa* PAO1, which produces Fluopsin C to cope with excess Cu stress.[93]

E. Bis-Thiosemicarbazones (atsm, gtsm)

Bis-thiosemicarbazones (bTSC) form strong Cu(II)-complexes at neutral pH (log K of the bTSC 3-ethoxy-2-oxobutyraldehyde-bis-thiosemicarbazone [compound called kts] was determined to be 19 at pH 7[94]). They form much stronger 1:1 complexes with Cu(II) than DTC or 8-HQ, even compared to the 1:2 complex of DTC and 8-HQ in the relevant concentration of low micromolar or below. This is mainly due to the chelate effect as they are tetradentate ligands (Figure 11). Using a drug screen assay, Speer *et al.*[90] demonstrated that bTSC induces a copper hypersensitivity phenotype in *M. tuberculosis* and proposed that copper complexation was the potential mechanism.

Cu-bTSC complexes are quite stable in biological environments. They form neutral, lipophilic compounds that enter the cell by passive membrane crossing. They are weaker Cu(I) than Cu(II) chelators but still have a log K of around 13 at pH 7 for Cu(I).[95] Once inside the cells, Cu(II) bound to bTCS could be reduced to Cu(I) and transferred from bTCS to GSH. However, depending on the complex formed and on its redox potential, such reduction can be very slow (Table 3, Figure 11). For instance, the Cu(II)-atsm complex has been shown to not be reduced *in vitro* even

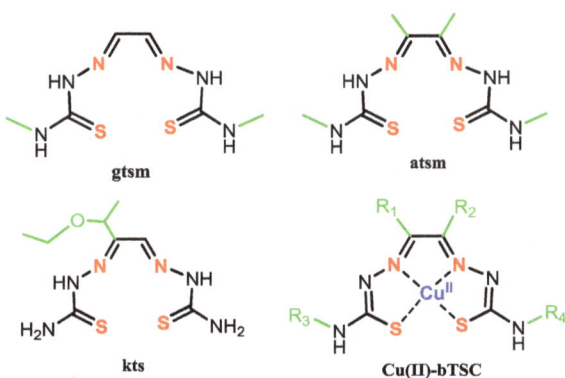

Figure 11. Structure of bis-thiosemicarbazones (bTSC) and their Cu(II)-complexes. Structural difference between the ligands are highlighted in green.

in the presence of millimolar of GSH, in line with Cu(II)-atsm low redox potential (see Table 3, Figure 11).[72,95] In *E. coli*, Cu seems to be released from Cu-atsm, but with a much slower rate than from Cu-gtsm.[91] The difference observed between Cu(II)-atsm and Cu(II)-gtsm in terms of redox potential and reactivity is reflected in term of level of toxicity. Cu(II)-gtsm exerts a higher level of toxicity on bacteria than Cu(II)-atsm. This difference can be explained by the fact that Cu-gtsm most likely acts as an ionophore (mechanism I): it enters the cells and Cu is rapidly released by the action of intracellular thiols (GSH and others), the fast accumulation of intracellular Cu(I) exerting toxicity. In contrast, Cu(II)-atsm is less toxic, which is consistent with the fact that Cu(II)-atsm is more stable, releases Cu(I) only very slowly and this gives more time to the bacteria to react and to switch on the defense system.[91] Toxicity of cytoplasmic Cu(I) released from Cu(II)-gtsm was further confirmed by the increased sensitivity of *E. coli* strain deleted of *copA* gene, encoding the exporter which removes Cu cytoplasmic excess. In contrast, the deletion of *cueO* gene, encoding a periplasmic copper oxidizing enzyme (Figure 3), has no effect on cell survival toward Cu(II)-gtsm.[91] This finding suggests that dissociation of Cu(II)-gtsm occurs specifically in the cytosol.

A recent study from Totten *et al.*[80] defined the impact of different compounds toward *Mycoplasma* and *Ureaplasma* species. In their assays, they observed a bactericidal activity of gtsm compound, however, not in a

Table 3. Thermodynamic and kinetics properties of Cu-L complexes.

Compound class	Representative	Affinity (pH 7.4) in log Kd for Cu(II)	Cu(I)	E° Cu(II)/Cu(I) (V/NHE)	Intracellular reactivity (cytoplasm)	Denticity
Cu alone				0.16	– Fast reduction of Cu(II) – Cu(I) binds to GSH	—
Dithiocarbamate (DTC)	DETC Cu + DETC Cu-DETC + DETC	11.7[c] 12.2[c]		-0.14[k]	– Fast reduction – Complete dissociation – Cu(I) binds to GSH	2
8-hydroxyquinoline (HQ)	Clioquinol (CQ) Cu + CQ Cu + HQ Cu-CQ + CQ Cu-HQ + HQ	10.8[f] 9.75[g] 9.2[f] 8.65[g]		-0.36[h]	– Fast reduction – Dissociation – Cu(I) binds to GSH	2
Phenanthrolines	Cu + Phen Cu-Phen + Phen Cu + 2 Phen BCS Cu + 2BCS	9.1[l] 6.8[l] 15.9[l] 12.5[d]	10.3[l] 5.5[l] 15.8[l] 20.8[d]	0.17[i] 0.62[j]	– Fast reduction – Partial dissociation – Cu(I) binds to GSH	2
Pyrithione	Cu + PT Cu(II)-PT + PT	> 8.5[e] 5.9[e]			– Fast reduction – Dissociation – Cu(I) binds to GSH	2
Bis-thiosemicarbazone	Cu + Atsm Cu + Gtsm Cu + Kts	~13[a] ~13[a] ~19[b]		-0.40[a] -0.24[a] -0.12[b] (pH 6.6)	– Stable – Slow – Dissociation	4

a 95, b. 94, c in methanol;[97] d 98, e 99, f in 80% Methanol and 20% H$_2$O, [100] g 101, h 102, i 103, j 104, k 105, l 106 copper dependent manner. Still, the use of Cu(II) chelator in the media decreased gtsm antibacterial activity, in line with the fact that small amount of Cu in the media is sufficient to boost the growth inhibitory effect of gtsm.

The higher toxicity of Cu-gtsm compare to Cu-atsm applies to other bacteria (Gram-positive and -negative).[91,96] However, the susceptibility of the bacteria is very different, and is directly related to the bacterial physiology. Bacteria with a high developed Cu-detoxification system like *E. coli* are quite resistant, other with a poorer Cu extrusion system, like *N. gonorrhoeae* are much more susceptible (IC 50 < μM). This can be explained by the fact that *N. gonorrhoeae* have only one exporter, CopA, whose expression is not inducible by Cu. Moreover, in *N. gonorrhoeae* an additional toxicity mechanism was described related to the inhibition of dehydrogenases in the respiratory chain.[96] The results suggested that Cu-bTSC complexes bind to membrane proteins, likely in hydrophobic pockets where they interfere with ubiquinol. This mechanism is independent of Cu release or redox activity, in line with experiments showing that Zn-bTSC showed similar activity.[96] However, other mechanisms may explain the toxicity of Cu-gtsm, as bacteria growing under fermentation (*S. pneumoniae* and *E. coli*) are also highly sensitive to Cu-gtsm.[91]

X. Peptides-Based Cu-Chelators

Antimicrobial peptides (AMPs) are described as peptides of about 10 to 100 amino acids that can be either naturally produced by the immune system to combat invading pathogens or chemically synthesized with non-natural sequences. For certain natural AMPs, the binding of copper or other metal ions was proposed to be required for their activity. The main idea was that Cu-binding to AMPs may improve the antimicrobial activity either through Cu-catalyzed ROS production or Cu-induced conformational changes enhancing the interaction of the AMP with its target. Two main approaches are reported: (i) identification of native AMPs containing a high affinity-Cu-binding site, (ii) engineering such a site in a native or artificial peptide (recent review on AMPs[107]).

Short peptides are often dynamic molecules lacking a well-defined 3D structure. Hence, Cu-binding to peptides is generally weaker than to

XZH (ATCUN)   **XH**

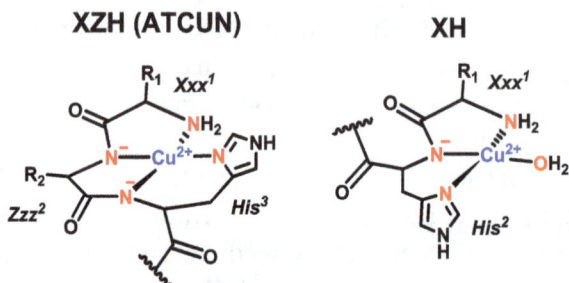

Figure 12.   Structure of Cu(II)-XZH and Cu(II)-XH complexes.

Cu-proteins with defined 3D-structures, due to entropic penalty for arranging the peptidic ligands in a constrained coordination sphere around the metal ion. To date, no simple Cu(I)-binding peptide motifs with sub-femtomolar affinity (unlike Cu(I)-proteins) are known. Instead, two naturally occurring His-containing Cu(II)-binding peptide motifs can attain a sub-picomolar affinity at pH 7.4. They require a His residue at either position 2 or 3 in a peptide with a free N-terminus, *i.e.*, NH$_2$-Xxx-His-Peptide (XH) or NH$_2$-Xxx-Zzz-His-Peptide (XZH, also known as ATCUN motif)[73] (Figure 12). Most studies focused on these motifs as they show comparable affinity to endogenous Cu(II)-carriers (such as serum albumin in human blood), which makes them apt to bind Cu(II) in biological environments. Antimicrobial activities of several native AMPs containing a XZH motif have been compared with and without Cu (reviewed in[108]). Generally, the impact of Cu was quite low, with a similar antimicrobial activity (measured as MIC values) whether Cu was added to AMPs or not. The modest Cu effect could be explained by the redox properties of Cu bound to XZH which induces a very low ROS production.[109] However, ATCUN peptides bearing an additional Cu(I)-stabilizing site show higher antimicrobial activity. For instance, the salivary peptide Histatin-5 (Hist-5) shows antimicrobial activity against the yeast *C. albicans* and *P. gingivalis*[108] (Table 4). Hist-5 binds Cu(II) in a XZH motif, but has in addition a vicinal Cu(I)-binding *bis*-His (-HH-) motif which might favor the redox cycling and hence the ROS production. Remarkably, the mutation of Cu-binding His residues from the HH motif

Table 4. Peptides mentioned in this chapter and their Cu-binding motifs (italic for Cu(II), underlined for Cu(I), and bold for both).

PEPTIDE	Cu binding motifs
Hist-5	*DSH*AKR<u>HH</u>...
Piscidin 1	*FF***H**<u>H</u>...
Piscidin 3	*FI***H**<u>H</u>...
Ixosin	*GL***H**KV<u>M</u>...

significantly affects the antimicrobial activity of Hist-5.[110,111]. Likewise, ATCUN-bearing AMPs such as Piscidins and Ixosin could show higher ROS production than simple ATCUN peptides thanks to their Cu(I)-binding motifs, *i.e.*, -HH- and -HXZM-, respectively[108] (Table 4). Nevertheless, the impact of such Cu(I)-binding sites on the ROS production by ATCUN-AMPs has never been assessed quantitatively.[112,113] On the other hand, in the cytosol, Cu(II)-XZH complexes are slowly reduced and dissociated by GSH.[114,115] In this case, a Cu(I)-binding site could lead to a faster dissociation of Cu from Cu(II)-XZH, *via* the faster reduction to Cu(I) and subsequent dissociation. Overall, this implies that (i) the potential pro-oxidant activity might take place only in the extracellular or periplasmic compartments (mechanism II) and (ii) these complexes might act as ionophores (mechanism I), even though no report has yet shown any AMP with an efficient ionophore activity.

Recently, the shorter XH motif in an AMP peptide (XH-AMP) was investigated and compared to XZH-AMP and AMP only. Cu(II)-XH is slightly more active in ROS production than Cu(II)-XZH, although still less compared to Cu in buffer or in a well redox active Cu-complex such as Cu-Phen$_2$.[72] Nevertheless, the same antimicrobial activity as pure AMP against *E. coli* was observed with Cu-bound or unbound XH- or XZH-AMP.[116]

Besides, structural changes upon Cu-binding to XH and XZH motifs are likely not drastic as these motifs are short and located at the N-terminus.

XI. Recent Advances Toward the Future Use of Copper in Medicine

Copper is a potent antimicrobial agent and even though used since thousands of years, it regains interest as a promising compound to increase therapeutic efficiency toward recalcitrant microbial infections. Indeed, as increasing amount of copper in bacteria environment leads to cell death, such property is even used by macrophages to kill invading pathogens. Cu-mechanism of action to explain toxicity as well as strategies developed by cells to survive such thread start to be better understood and these fundamental knowledges will pave the way toward the discovery of copper-related compounds with application in medicine. Copper-complexes, as described in this chapter, show interesting properties demonstrating their potential for clinical treatment. They present a higher level of bactericide activity compared to Cu or the ligand alone. For most of them, Cu-L permit to cross bacteria membranes with a higher efficiency than "free" Cu and once in the reducing cytoplasmic compartment they dissociate and the released Cu accumulates and perturbs intracellular pathways. This ionophore activity has been mainly described for most of the Cu-complexes. Other mechanisms of action could occur, however, such as a gain of function to induce ROS production by the Cu-L once inside the periplasmic compartment. This requires further experimental evidences to be confirmed.

Cu-gtsm is a promising Cu-complex, highly effective toward multidrug-resistant bacteria, giving some hope to cure disease related to highly resistant strains. Very interestingly, antimicrobial Cu-gtsm doses used to kill the extracellular mucosal pathogen *N. gonorrhoeae* did not show significant toxicity toward cervical epithelial cells.[91] Recently, other molecules, like the 1-hydroxy-5-*R*-pyridine-2(1*H*)-thiones (with *R*: COOMe, COOEt, CF$_3$), have been shown to efficiently inhibit *M. tuberculosis* at very low amount and in a copper-dependent manner.[117]

However, the next challenge in this area of study will be to develop specific antibacterial compounds to target pathogenic bacteria but neither host cells nor non-pathogenic bacteria. Even if some complexes might be interesting to further study for clinical applications, most of the Cu-complexes described in this issue have no selectivity toward bacteria *versus* mammalian cells. Some of these compounds are even used to treat cancer and are known to negatively impact eukaryotic cells.[118,119] To increase selectivity, several strategies can be envisioned.

Festa *et al.*[83] used a compound that gets activated once inside macrophages. They took advantage of the high level of ROS as well as of copper in macrophages to transform a non-toxic compound into a highly toxic one. In particular, oxidation of a pro-drug released the 8-HQ ligand that complexed Cu and then acted as ionophore to aid the immune system to kill the pathogens. They were able to show that such conditional activation worked *in vitro* and in mouse model on the pathogen *C. neoformans.*[83]

Another strategy followed by Wolschendorf and coworkers was to screen different compounds in combination with copper against *S. aureus.*[120] They were able to highlight the potential of new copper-dependent molecules such as thiourea inhibitors which were able to inhibit efficiently *S. aureus* growth and seemed tolerated in cell culture. They also demonstrated the toxicity of PZP-915 (5-Benzyl-3-(4-chlorophenyl) -2-methyl-4H,7H-pyrazolo [1,5-a] pyrimidin-7-one) toward *S. aureus* resistant strains with a certain selectivity toward bacteria over eukaryotic cells.[121]

The use of peptide with Cu-binding motifs is another way to obtain higher selectivity toward bacteria as it is exemplified by the natural AMPs. Peptides could also be used for targeting a Cu-site to compartments or proteins. Accordingly, cytoplasm-targeted peptides could act as ionophores by adding a Cu(II)-site to the peptide. This strategy would need optimization of the peptide sequence toward cell-penetration, so far, an area of research poorly explored. However, it is not clear whether peptides can be as efficient as small organic ligands. For the antimicrobial strategy consisting in employing a Cu-site on an AMP to catalyze ROS production, reports are mainly limited to Xxx-Zzz-His motifs. This motif is quite redox inert as it stabilizes Cu(II) and hence ROS production is low. Other high affinity Cu-binding sites are needed to increase ROS production. This seems to be very challenging for short peptides, and protein-like folding and/or organic ligands are needed to reach the balance between stability and redox reactivity. Moreover, peptides form a scaffold which can be conjugate to other entities of interest, such as fluorophore for tracking other antimicrobial molecules.

Synergic effect of copper/copper derivatives with antibiotics is also an alternative way of finding highly specific molecules toward pathogenic bacteria.[55] ATP-6K in presence of copper and in combination with ampicillin was shown to restore ampicillin antimicrobial activity against resistant strains without affecting eukaryotic cells.[122]

These strategies among others should allow to overcome the limitation of copper use in medical applications. Indeed, study on copper and its derivatives is of high interest to develop novel efficient antimicrobial agents. However, the use of metals to kill pathogenic bacteria has to be carefully thought. Indeed, feeding animals by copper in industry lead to the spread of resistance mechanism such as the *pco* genomic island in *E. coli*.[123] as well as of a functional *cus* locus in *S. enterica* that increases its resistance toward metal.[124] This raises the concern of the presence of metal in food which acts as a disseminating agent of resistance mechanisms.

The development of Cu-dependent inhibitors might impose selection pressure to bacteria and lead to the development of metal resistance strategies. Efficient copper pumps, for instance, could help bacteria to resist Cu-complexes ionophore activity. However, as described in this review, copper induces bacteria cell death by targeting a variety of cellular processes. These multiple targets make Cu-dependent inhibitors very attractive compared to some antibiotics which target a single process that bacteria can easily overcome. A good balance is required: the use of small amount of copper specifically dedicated to kill invading microorganisms should be clearly envisioned whiles decreasing the excessive use of copper as food additive and biocide.

XII. References

1. Decaria, L.; Bertini, I.; Williams, R. J. P. *Metallomics.* **2011**, *3*(1), 56–60. doi: 10.1039/C0MT00045K.
2. Dupont, C. L.; Grass, G.; etRensing, C. *Metallomics.* **2011**, *3*(11), 1109. doi: 10.1039/c1mt00107h.
3. Hong Enriquez R. P.; Do, T. N. *Life.* **2012**, *2*(4), 274–285. doi: 10.3390/life2040274.
4. Bertini, I.; Cavallaro, G.; McGreevy, K. S. *Coord. Chem. Rev.* **2010**, *254*(5–6), 506–524. doi: 10.1016/j.ccr.2009.07.024.

5. Borkow, G.; Gabbay, J. *CMC* **2005**, *12*(18), 2163–2175. doi: 10.2174/0929867054637617.

6. Borkow, G.; Gabbay, J. *Curr. Chem. Biol.* **2009**, *3*(3), 272–278. doi: 10.2174/187231309789054887.

7. Grass, G.; Rensing, C.; Solioz, M. *Appl. Environ. Mictobiol.* **2011**, *77*(5), 5. doi: 10.1128/AEM.02766-10.

8. Arendsen, L. P.; Thakar, R.; Sultan, A. H. *Clin. Microbiol. Rev.* **2019**, *32*(4), e00125-18. /cmr/32/4/CMR.00125-18.atom. doi: 10.1128/CMR.00125-18.

9. Hodgkinson, V.; Petris, M. J. *J. Biol. Chem.* **2012**, *287*(17), 13549–13555. doi: 10.1074/jbc.R111.316406.

10. Hood, M. I.; Skaar, E. P. *Nat. Rev. Microbiol.* **2012**, *10*(8), 525–537. doi: 10.1038/nrmicro2836.

11. Wagner, D. *et al.* *J. Immunol.* **2005**, *174*(3), 1491–1500. doi: 10.4049/jimmunol.174.3.1491.

12. Djoko, K. Y.; Ong, C. Y.; Walker, M. J.; McEwan, A. G. *J. Biol. Chem.* **2015**, *290*(31), 31. doi: 10.1074/jbc.R115.647099.

13. White, C.; Lee, J.; Kambe, T.; Fritsche, K.; Petris, M. J. *J. Biol. Chem.* **2009**, *284*(49), 49. doi: 10.1074/jbc.M109.070201.

14. Giachino, A.; Waldron, K. *J. Mol. Microbiol.* **2020**, Sep; *114*(3):377–390. doi: 10.1111/mmi.14522.

15. G. Ghssein *et al.*, *Science*, **2016**, *352*(6289) 1105–1109.

16. H. J. Kim *et al.*, *Science*, **2004**, *305*(5690), 1612–1615, doi: 10.1126/science.1098322.

17. Koh, E.-I.; Robinson, A. E.; Bandara, N.; Rogers, B. E.; Henderson, J. P. *Nat. Chem. Biol.* **2017**, *13*(9), 1016–1021. doi: 10.1038/nchembio.2441.

18. Andrei, A. *et al. Membranes.* **2020**, *10*(9), 242. doi: 10.3390/membranes10090242.

19. Rosenzweig, A. C. *Chem. Biol.* **2002**, *9*(6), 673–677. doi: 10.1016/S1074-5521(02)00156-4.

20. Mealman, T. D.; Blackburn, N. J.; McEvoy, M. M. in *Current Topics in Membranes.* **2012**, *69*, Elsevier, pp. 163–196. doi: 10.1016/B978-0-12-394390-3.00007-0.

21. Munson, G. P.; Lam, D. L.; Outten, F. W.; O'Halloran, T. V. *J. Bacteriol.* **2000**, *182*(20), 5864–5871. doi: 10.1128/JB.182.20.5864-5871.2000.

22. Outten, F. W.; Huffman, D. L.; Hale, J. A.; O'Halloran, T. V. *J. Biol. Chem.* **2001**, *276*(33),33. doi: 10.1074/jbc.M104122200.

23. Singh, S. K.; Grass, G.; Rensing, C.; Montfort, W. R. *J. Bacteriol.* **2004**, *186*(22), 22. doi: 10.1128/JB.186.22.7815-7817.2004.

24. Vita, N. *et al. Nature.* **2015**, *525*(7567), 140–143. doi: 10.1038/nature14854.

25. Vita, N. *et al. Sci. Rep.* **2016**, *6*(1), 39065. doi: 10.1038/srep39065.

26. Stewart, L. J. *et al. mBio.* **2020**, *11*(6), e02804-20. /mbio/11/6/mBio.02804-20.atom. doi: 10.1128/mBio.02804-20.

27. Durand, A. *et al. Metallomics.* **2021**, *13*(12), mfab067. doi: 10.1093/mtomcs/mfab067.

28. Durand, A. *et al. mBio.* **2015**, *6*(5), e01007–15. doi: 10.1128/mBio.01007-15.
29. Hausrath, A. C.; Ramirez, N. A.; Ly, A. T.; McEvoy, M. M. *J. Biol. Chem.* **2020**, *295*(32), 11364–11376. doi: 10.1074/jbc.RA120.013907.
30. Rensing, C.; Grass, G. *FEMS Microbiol Rev.* **2003**, *27*(2–3), 197–213. doi: 10.1016/S0168-6445(03)00049-4.
31. Nunes, I. *et al. FEMS Microbiol. Ecol.* **2016**, *92*(11), fiw175. doi: 10.1093/femsec/fiw175.
32. Rivera-Millot A. *et al., Commun. Biol.* **2021**, *4*(1), 46, doi: 10.1038/s42003-020-01580-2.
33. Barahona, S.; Castro-Severyn, J.; Dorador, C.; Saavedra, C.; Remonsellez, F. *Genes* **2020**, *11*(8), 844. doi: 10.3390/genes11080844.
34. Navarro, C. A.; von Bernath, D.; et Jerez, C. A. *Biol. Res.* **2013**, *46*(4), 363–371. doi: 10.4067/S0716-97602013000400008.
35. Macomber, L.; Rensing, C.; Imlay, J. A. *J. Bacteriol.* **2007**, *189*(5), 5. doi: 10.1128/JB.01357-06.
36. Navarro, C. A.; Orellana, L. H.; Mauriaca, C.; Jerez, C. A. *Appl. Environ. Microbiol.* **2009**, *75*(19), 6102–6109. doi: 10.1128/AEM.00308-09.
37. Davis, A. V.; O'Halloran, T. V. *Nat. Chem. Biol.* **2008**, *4*(3), 148–151. doi: 10.1038/nchembio0308-148.
38. Haas, K. L.; Franz, K. J. *Chem. Rev.* **2009**, *109*(10), 4921–4960. doi: 10.1021/cr900134a.
39. Irving, H.; Williams, R. J. P. *J. Chem. Soc.* **1953**, 3192. doi: 10.1039/jr9530003192.
40. Lisher, J. P.; Giedroc, D. P. *Front. Cell. Infect. Microbiol.* **2013**, *3*. doi: 10.3389/fcimb.2013.00091.
41. Halliwell, B.; Gutteridge, J. M. C. *Biochem. J.* **1984**, *219*(1), 1–14. doi: 10.1042/bj2190001.
42. Morgan, M. T.; Nguyen, L. A. H.; Hancock, H. L.; Fahrni, C. J. *J. Biol. Chem.* **2017**, *292*(52), 21558–21567. doi: 10.1074/jbc.M117.817452.
43. Changela, A. *Science* **2003**, *301*(5638), 1383–1387. doi: 10.1126/science.1085950.
44. Outten, C. E. *Science* **2001**, *292*(5526), 2488–2492. doi: 10.1126/science.1060331.
45. Macomber, L.; Imlay, J. A.; *Proc. Natl. Acad. Sci.* **2009**, *106*(20), 8344–8349. doi: 10.1073/pnas.0812808106.
46. Cong, X.; Poyton, M. F.; Baxter, A. J.; Pullanchery, S.; Cremer, P. S. *J. Am. Chem. Soc.* **2015**, *137*(24), 7785–7792. doi: 10.1021/jacs.5b03313.
47. Poyton, M. F.; Sendecki, A. M.; Cong, X.; Cremer, P. S. *J. Am. Chem. Soc.* **2016**, *138*(5), 1584–1590. doi: 10.1021/jacs.5b11561.
48. Hong, R.; Kang, T. Y.; Michels, C. A.; Gadura, N. *Appl. Environ. Microbiol.* **2012**, *78*(6), 1776–1784. doi: 10.1128/AEM.07068-11.
49. May, K. L.; Lehman, K. M.; Mitchell, A. M.; Grabowicz, M. *mBio* **2019**, *10*(3), e00618–19. /mbio/10/3/mBio.00618-19.atom. doi: 10.1128/mBio.00618-19.
50. Peters, K. *et al. Proc. Natl. Acad. Sci. USA.* **2018**, *115*(42), 10786–10791. doi: 10.1073/pnas.1809285115.

51. Eichhorn, G. L.; Shin, Y. A. *J. Am. Chem. Soc.* **1968**, *90*(26), 7323–7328. doi: 10.1021/ja01028a024.
52. Geierstanger, B. H.; Kagawa, T. F.; Chen, S. L.; Quigley, G. J.; Ho, P. S. *J. Biol. Chem.* **1991**, *266*(30), 20185–20191. doi: 10.2210/pdb1d40/pdb.
53. Rifkind, J. M.; Shin, Y. A.; Heim, J. M.; Eichhorn, G. L. *Biopolymers* **1976**, *15*(10), 1879–1902. doi: 10.1002/bip.1976.360151002.
54. Sagripanti, J. L.; Kraemer, K. H. *J. Biol. Chem.* **1989**, *264*(3), 1729–1734.
55. Dalecki, A. G.; Crawford, C. L.; Wolschendorf, F. in *Advances in Microbial Physiology.* **2017**, *70*, Elsevier, pp. 193–260. doi: 10.1016/bs.ampbs.2017.01.007.
56. Manzanillo, P. S.; Shiloh, M. U.; Portnoy, D. A.; Cox, J. S. *Cell Host & Microbe* **2012**, *11*(5), 469–480. doi: 10.1016/j.chom.2012.03.007.
57. Whitchurch, C. B.; *Science* **2002**, *295*(5559), 1487–1487. doi: 10.1126/science.295.5559.1487.
58. Hiniker, A.; Collet, J.-F.; Bardwell, J. C. A. *J. Biol. Chem.* **2005**, *280*(40), 33785–33791. doi: 10.1074/jbc.M505742200.
59. Zuily, L. *et al. mBio* **2022**, *13*(2), e03251–21. doi: 10.1128/mbio.03251-21.
60. Vergnes A. *et al.*, *PLoS Genet* **2022**, *18*(7), e1010180, doi: 10.1371/journal.pgen.1010180.
61. Saporito-Magriñá, C. M. *et al. Metallomics* **2018**, *10*(12), 1743–1754. doi: 10.1039/C8MT00182K.
62. Wiebelhaus, N.; Zaengle-Barone, J. M.; Hwang, K. K.; Franz, K. J.; Fitzgerald, M. C. *ACS Chem. Biol.* **2021**, *16*(1), 214–224. doi: 10.1021/acschembio.0c00900.
63. Foster, A. W.; Osman, D.; Robinson, N. J. *J. Biol. Chem.* **2014**, *289*(41), 28095–28103. doi: 10.1074/jbc.R114.588145.
64. Tottey, S. *et al. Nature* **2008**, *455*(7216), 1138–1142. doi: 10.1038/nature07340.
65. Chillappagari, S.; Seubert, A.; Trip, H.; Kuipers, O. P.; Marahiel, M. A.; Miethke, M. *J. Bacteriol.* **2010**, *192*(10), 2512–2524. doi: 10.1128/JB.00058-10.
66. Djoko, K. Y.; McEwan, A. G. *ACS Chem. Biol.* **2013**, *8*(10), 2217–2223. doi: 10.1021/cb4002443.
67. Johnson, M. D. L.; Kehl-Fie, T. E.; Rosch, J. W. *Metallomics* **2015**, *7*(5), 786–794. doi: 10.1039/C5MT00011D.
68. Tan, G. *et al. Appl. Environ. Microbiol.* **2017**, *83*(16), 16. doi: 10.1128/AEM.00867-17.
69. Gugala, N.; Salazar-Alemán, D. A.; Chua, G.; Turner, R. J. *Metallomics* **2022**, *14*(1), mfab071. doi: 10.1093/mtomcs/mfab071.
70. O'Hern C.; Djoko, K. Y. MBio, **2022**, *13*(3), e00044-22, doi:10.1128/mbio.00434-22
71. Bossak-Ahmad, K.; Frączyk, T.; Bal, W.; Drew, S. C. *ChemBioChem* **2020**, *21*(3), 331–334. doi: 10.1002/cbic.201900435.
72. Santoro, A.; Calvo, J. S.; Peris-Díaz, M. D.; Krężel, A.; Meloni, G.; Faller, P. *Angew. Chem. Int. Ed.* **2020**, *59*(20), 7830–7835. doi: 10.1002/anie.201916316.
73. Gonzalez, P. *et al. Chem. Eur. J.* **2018**, *24*(32), 8029–8041. doi: 10.1002/chem.201705398.

74. Rajalakshmi, S.; Fathima, A.; Rao, J. R.; Nair, B. U. *RSC Adv.* **2014**, *4*(60), 32004–32012. doi: 10.1039/C4RA03241A.
75. Phillips, M.; Malloy, G.; Nedunchezian, D.; Lukrec, A.; Howard, R. G. *Antimicrob. Agents Chemother.* **1991**, *35*(4), 785–787. doi: 10.1128/aac.35.4.785.
76. Byrne, S. T. *et al. Antimicrob. Agents Chemother.* **2007**, *51*(12), 4495–4497. doi: 10.1128/AAC.00753-07.
77. Zaldívar-Machorro, V. J.; López-Ortiz, M.; Demare, P.; Regla, I.; Muñoz-Clares, R. A. *Biochimie* **2011**, *93*(2), 286–295. doi: 10.1016/j.biochi.2010.09.022.
78. Dalecki, A. G. *et al. Antimicrob Agents Chemother.* **2015**, *59*(8), 8. doi: 10.1128/AAC.00692-15.
79. Menghani, S. V. *et al. Microbiol Spectr.* **2021**. doi: 10.1128/Spectrum.00778-21.
80. Totten, A. H.; Crawford, C. L.; Dalecki, A. G.; Xiao, L.; Wolschendorf, F.; Atkinson, T. P. *Front. Microbiol.* **2019**, *10*, 1720. doi: 10.3389/fmicb.2019.01720.
81. Skrott, Z. *et al. Nature* **2017**, *552*(7684), 194–199. doi: 10.1038/nature25016.
82. Xu, L. *et al. Angew. Chem. Int. Ed.* **2019**, *58*(18), 6070–6073. doi: 10.1002/anie.201814519.
83. Festa, R. A.; Helsel, M. E.; Franz, K. J.; Thiele, D. J. *Chem. Biol.* **2014**, *21*(8),8. doi: 10.1016/j.chembiol.2014.06.009.
84. Tardito, S. *et al. J. Am. Chem. Soc.* **2011**, *133*(16), 6235–6242. doi: 10.1021/ja109413c.
85. Shah, S. *et al. Antimicrob. Agents Chemother.* **2016**, *60*(10), 5765–5776. doi: 10.1128/AAC.00325-16.
86. Cahoon, L. *Nat Med.* avr. **2009**, *15*(4), 356–359. doi: 10.1038/nm0409-356.
87. Ananthan, S. *et al. Tuberculosis.* **2009**, *89*(5), 334–353. doi: 10.1016/j.tube.2009.05.008.
88. Gilbert, B. C.; Silvester, S.; Walton, P. H. *J. Chem. Soc., Perkin Trans.* **1999**, *2*(6), 1115–1122. doi: 10.1039/a901179j.
89. Calvo, J. S.; Lopez, V. M.; Meloni, G. *Metallomics* **2018**, *10*(12), 1777–1791. doi: 10.1039/C8MT00264A.
90. Speer, A. *et al. Antimicrob. Agents Chemother.* **2013**, *57*(2), 1089–1091. doi: 10.1128/AAC.01781-12.
91. Djoko, K. Y.; Goytia, M. M.; Donnelly, P. S.; Schembri, M. A.; Shafer, W. M.; McEwan, A. G. *Antimicrob. Agents Chemother.* **2015**, *59*(10),10. doi: 10.1128/AAC.01289-15.
92. Chiem, K. *et al. Antimicrob. Agents Chemother.* **2015**, *59*(9), 5851–5853. doi: 10.1128/AAC.01106-15.
93. Patteson, J. B. *et al. Science* **2021**, *374*(6570), 1005–1009. doi: 10.1126/science.abj6749.
94. Petering, D. H. *Bioinorg. Chem.* **1972**, *1*(4), 273–288. doi: 10.1016/S0006-3061(00)81002-9.
95. Xiao, Z.; Donnelly, P. S.; Zimmermann, M.; Wedd, A. G. *Inorg. Chem.* **2008**, *47*(10), 4338–4347. doi: 10.1021/ic702440e.

96. Djoko, K. Y.; Paterson, B. M.; Donnelly, P. S.; McEwan, A. G. *Metallomics* **2014**, *6*(4), 854–863. doi: 10.1039/C3MT00348E.

97. Labuda, J.; Skatulokova, M.; Nemeth, M.; Gergely, S. *Chem. Zvesti.* **1984**, 597—605.

98. Bagchi, P.; Morgan, M. T.; Bacsa, J.; Fahrni, C. J. *J. Am. Chem. Soc.* **2013**, *135*(49), 18549–18559. doi: 10.1021/ja408827d.

99. Sun, P. J. Quintus, Fernando, Henry, Freiser, *Anal. Chem.* **1964**, *36*(13), 2485–2488. doi: 10.1021/ac60219a034.

100. Budimir, A.; Humbert, N.; Elhabiri, M.; Osinska, I.; Biruš, M.; Albrecht-Gary, A.-M. *J. Inorg. Biochem.* **2011**, *105*(3), 490–496. doi: 10.1016/j.jinorgbio.2010.08.014.

101. Smith, R. M.; Martell, A. E. in *Critical Stability Constants*. Boston, MA: Springer US, **1989**. doi: 10.1007/978-1-4615-6764-6.

102. Wehbe, M. *et al.* *Drug Deliv. Transl. Res.* **2018**, *8*(1), 239–251. doi: 10.1007/s13346-017-0455-7.

103. Postnikova, G. B.; Shekhovtsova, E. A. *Biochemistry Moscow* **2016**, *81*(13), 1735–1753. doi: 10.1134/S0006297916130101.

104. Chen, D.; Darabedian, N.; Li, Z.; Kai, T.; Jiang, D.; Zhou, F. *Anal. Biochem.* **2016**, *497*, 27–35. doi: 10.1016/j.ab.2015.12.014.

105. Hendrickson, A. R.; Martin, R. L.; Rohde, N. M. *Inorg. Chem.* **1976**, *15*(9), 2115–2119. doi: 10.1021/ic50163a021.

106. McBryde, W. A. E. *Can. J. Biochem.* **1967**, 2093–2100.

107. Portelinha, J. *et al.* *Chem. Rev.* **2021**, acs.chemrev.0c00921. doi: 10.1021/acs.chemrev.0c00921.

108. Alexander, J. L.; Thompson, Z.; Cowan, J. A. *ACS Chem. Biol.* **2018**, *13*(4), 844–853. doi: 10.1021/acschembio.7b00989.

109. Santoro, A.; Walke, G.; Vileno, B.; Kulkarni, P. P.; Raibaut, L.; Faller, P. *Chem. Commun.* **2018**, *54*(84), 11945–11948. doi: 10.1039/C8CC06040A.

110. Conklin, S. E.; Bridgman, E. C.; Su, Q.; Riggs-Gelasco, P.; Haas, K. L.; Franz, K. J. *Biochemistry* **2017**, *56*(32), 4244–4255. doi: 10.1021/acs.biochem.7b00348.

111. Melino, S.; Santone, C.; Di Nardo, P.; Sarkar, B. *FEBS J.* **2014**, *281*(3), 657–672. doi: 10.1111/febs.12612.

112. Esmieu, C.; Ferrand, G.; Borghesani, V.; Hureau, C. *Chem. Eur. J.* **2021**, *27*(5), 1777–1786. doi: 10.1002/chem.202003949.

113. Schwab, S.; Shearer, J.; Conklin, S. E.; Alies, B.; Haas, K. L. *J. Inorg. Biochem.* **2016**, *158*, 70–76. doi: 10.1016/j.jinorgbio.2015.12.021.

114. Santoro, A.; Ewa Wezynfeld, N.; Vašák, M.; Bal, W.; Faller, P. *Chem. Commun.* **2017**, *53*(85), 11634–11637. doi: 10.1039/C7CC06802F.

115. Stefaniak, E.; Płonka, D.; Szczerba, P.; Wezynfeld, N. E.; Bal, W. *Inorg. Chem.* **2020**, *59*(7), 4186–4190. doi: 10.1021/acs.inorgchem.0c00427.

116. Bouraguba, M. *et al.* *J. Inorg. Biochem.* **2020**, *213*, 111255. doi: 10.1016/j.jinorgbio.2020.111255.

117. Salina, E. G. *et al.* *Metallomics* **2018**, *10*(7), 992–1002. doi: 10.1039/C8MT00067K.

118. Djoko, K. Y.; Donnelly, P. S.; McEwan, A. G. *Metallomics* **2014**, *6*(12), 2250–2259. doi: 10.1039/C4MT00226A.
119. Oliveri, V. *Coord. Chem. Rev.* **2020**, *422*, 213474. doi: 10.1016/j.ccr.2020.213474.
120. Dalecki, A. G.; Malalasekera, A. P.; Schaaf, K.; Kutsch, O.; Bossmann, S. H.; Wolschendorf, F. *Metallomics* *8*(4), 412–421. 2016. doi: 10.1039/C6MT00003G.
121. Crawford, C. L. *et al*. *Metallomics* **2019**, *11*(4), 784–798. doi: 10.1039/C8MT00316E.
122. Crawford, C. L.; Dalecki, A. G.; Perez, M. D.; Schaaf, K.; Wolschendorf, F.; Kutsch, O. *Sci Rep.* **2020**, *10*(1), 8955. doi: 10.1038/s41598-020-65978-y.
123. Fang, L. *et al*. *Sci Rep.* **2016**, *6*(1), 25312. doi: 10.1038/srep25312.
124. Arai, N. *et al*. *Antimicrob. Agents Chemother.* **2019**, *63*(9), e00429–19. /aac/63/9/AAC.00429-19.atom. doi: 10.1128/AAC.00429-19.

2 Transition State Analogue Molecules as Mechanistic Tools and Inhibitors for Tyrosinase

Clarisse Faure,* Amaury du Moulinet d'Hardemare,*
Hélène Jamet,* Catherine Belle,*, ‖ Elisabetta
Bergantino,[†] Luigi Bubacco,[†] Maurizio Benfatto,[‡]
A. Jalila Simaan[§] and Marius Réglier[§, ¶]

*University of Grenoble Alpes, CNRS-UGA UMR 5250, DCM,
 CS 40700, 38058 Grenoble Cedex 9, France
[†]Department of Biology, University of Padova, Via Ugo Bassi
 58b, 35131 Padova, Italy
[‡]Laboratori Nazionali di Frascati — INFN, Via E. Fermi 44,
 00044 Frascati, Italy
[§]Aix Marseille Université, CNRS, Centrale Marseille, iSm2,
 Marseille, France
‖catherine.belle@univ-grenoble-alpes.fr
¶marius.reglier@univ-amu.fr

List of Abbreviations

Ab	*Agaricus bisporus*
Ao	*Aspergillus oryzae*
AUS	Aureusidin synthase
Bm	*Bacillus megaterium*
CO	Catechol oxidase
DHI	Dihydroxyindole
DHICA	Dihydroxyindole carboxylic acid
DNA	Desoxyribonucleic acid
ESEEM	Electron spin echo envelope modulation
EXAFS	Extended X-Ray Absorption Fine Structure
H-BPEP	2,6-Bis[(bis(2-pyridylethyl)amino) methyl]-4-methylphenol
H-BPMEP	2-[(Bis(2-pyridylmethyl)amino)methyl]-6-[(bis (2-pyridylethyl)amino)methyl]-4-methylphenol
H-BPMP	2,6-Bis[(bis(2-pyridylmethyl)amino) methyl]-4-methylphenol
Hc	Hemocyanins
HOPNO	2-Hydroxypyridine *N*-oxide
Hs	*Homo sapiens*
HSPNO	2-Pyridinethiol *N*-oxide
HYSCORE	Hyperfine sublevel correlation
Ib	*Ipomoea batatas*
Jr	*Juglans regia*
KA	Kojic acid
L-DOPA	L-3,4-dihydroxyphenylalanine
Mim	L-mimosine
Mj	*Marsupenaeus japonicas*
MM	Molecular mechanics
Ms	*Manduca sexta*

MXAN	Minuit XANES
PTU	Phenylthiourea
QM	Quantum mechanics
ROS	Reactive oxygen species
Sc	*Streptomyces castaneoglobisporus*
SP	Square-based pyramid
TBP	Trigonal bipyramid
Thuj	Thujaplicin
Trop	Tropolone
TRP1	Tyrosinase related protein 1
TRP2	Tyrosinase related protein 2
TSA	Transition-state analogues
TSC	Thiosemicarbazone
TYR	Tyrosinase
XAS	X-ray absorption spectroscopy

I. Introduction

Tyrosinases (TYRs) are copper-containing redox enzymes present in prokaryotes and eukaryotes (plants, arthropods, fungi, and mammals) where they are involved in tissue pigmentation, wound healing, radiation protection, and primary immune response. Acting in phenol oxidation, TYRs belong to the class of polyphenol oxidases (PPOs). The TYR-catalyzed oxygenation consists of two consecutive reactions, namely the *ortho*-hydroxylation of phenol to catechol (cresolase or monophenolase activity, EC 1.14.18.1) and its subsequent oxidation into *ortho*-quinone (catecholase or diphenolase activity, EC 1.10.3.1) (Figure 1). In organisms, TYR is involved in the biosynthesis of melanin pigments.

II. Biological Functions of Melanins

Melanin pigments are amorphous polymers of high molecular weight.[1,2] Their structures remain poorly defined and differ according to the

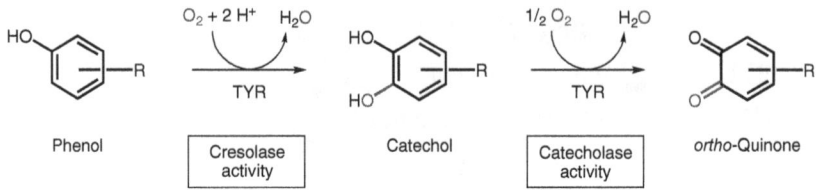

Figure 1.   TYR-catalyzed oxygenation reaction.

organisms. The common feature of all melanins is that they are formed by the polymerization of phenolic and indolic compounds catalyzed by TYR or TYR-like enzymes.

A. Melanin as Color Pigment

In humans, the color of the skin, hair, and eyes mainly depends on the concentration and ratio of eumelanin and pheomelanin. Human melanin is produced in melanosomes, which are organelles within cells called melanocytes found in the epidermis of the skin. Human melanocytes produce two chemically distinct types of melanin pigments, the brown-black colored pigment eumelanin, and the yellow-red colored pigment pheomelanin[3,4] (Figure 2). With broad optical absorption, eumelanin exhibits colors ranging from black to brown. Not only does eumelanin absorb all frequencies of visible light, but its optical absorption spectrum extends to the ultraviolet (UV) and infrared (IR) regions. With this characteristic, some consider eumelanin as "darker than black," a base for the *"outrenoir"* of the French painter Pierre Soulages. Other melanin pigments such as pheomelanin exhibit colors ranging from yellow to orange depending on their structures. The mixture of these melanin pigments is responsible for the coloring of the integuments in the animal kingdom.

The major role of melanins is to protect the skin from the harmful effects of UV rays and thus to prevent the development of skin cancers. In addition, eumelanin and pheomelanin play an important detoxification role within melanocytes and keratinocytes due to their ability to bind ions and various chemicals (Ca, Zn, Cu, Fe, orthoquinone).[5–7] Eumelanin is a

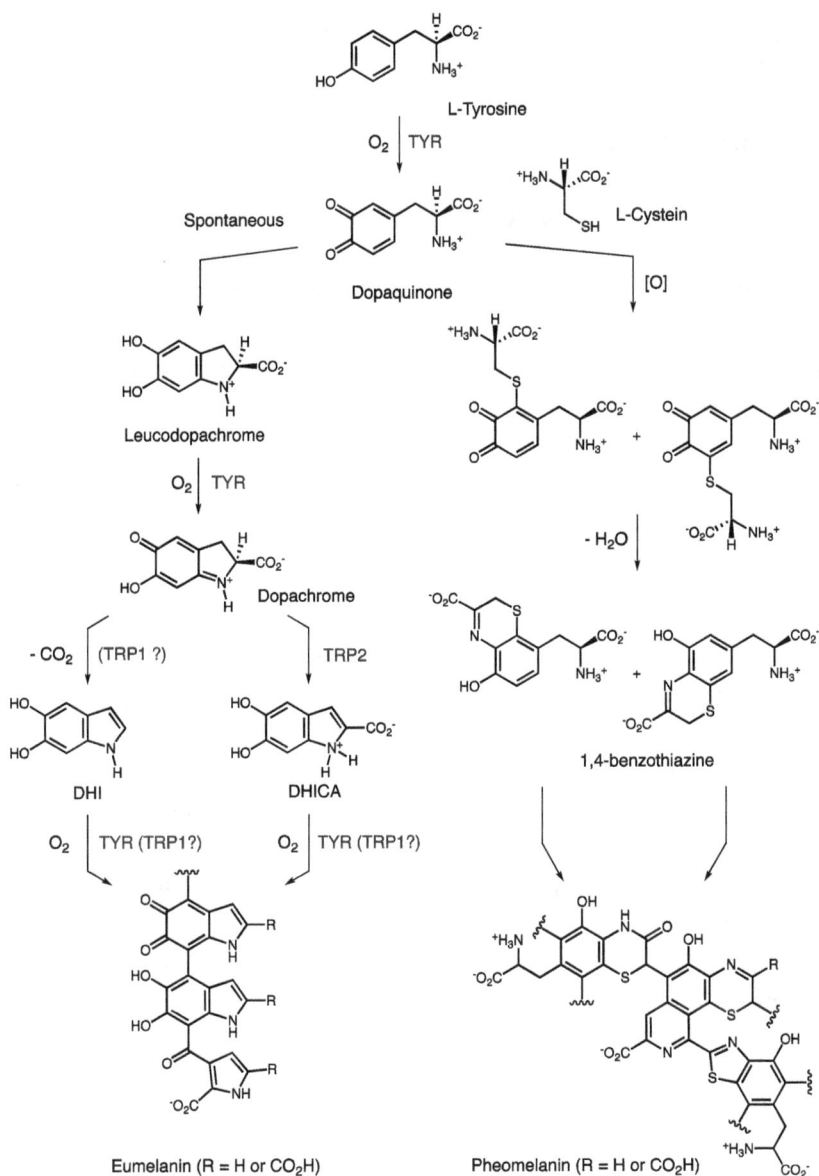

Figure 2. Eumelanin and pheomelanin biosynthesis from L-tyrosine (L-Tyr).

highly heterogeneous copolymer consisting of dihydroxyindole (DHI) and dihydroxyindole carboxylic acid (DHICA) units in reduced or oxidized form. Pheomelanin is mainly composed of benzothiazine derivatives. The first steps in eumelanin and pheomelanin biosynthesis are catalyzed by TYR, the hydroxylation of L-tyrosine into L-DOPA (monophenolase or cresolase activity) followed by its subsequent oxidation into dopaquinone (diphenolase or catecholase activity). The pheomelanin pigments are the result of a cascade of non-catalyzed reactions, which include the nucleophilic addition of a L-cysteinate onto dopaquinone followed by a heterocyclization into 1,4-benzothiazine and polymerization (Figure 2).

Many animals produce melanin, such as birds (feather coloring) and some protozoa. Melanin is also found as a pigment in fungi (mycete) in which it colors and strengthens the cell wall.

B. Melanin as a Defensive Barrier

In all organisms, melanins act as protective pigments against stresses that involve cellular damage such as sun UV rays and reactive oxygen species (ROS). Melanins are also known to protect against ionizing radiations. In radiotrophic fungi, melanin has a protecting role against γ rays and at the same time the received energy is harnessed for the growth of fungi. Melanin can also act as a defense against pathogen attack. In invertebrates, the innate immune defense system against invading pathogens involves melanin. Within minutes after infection, the pathogen is encapsulated in a cocoon of melanin and the generation of ROS during the formation of this cocoon leads to the destruction of the pathogen. Finally, melanin is found in the ink used by many cephalopods as a defense mechanism against predators.

In addition to these defense-related roles, melanin can be used by some pathogens as an attack agent. It is indeed reported that some human and plant pathogenic microorganisms, such as *Cryptococcus neoformans* (fungus), melanin plays an important role in the virulence and pathogenicity of the microorganism by protecting it against the immune responses of the host.[8]

III. Tyrosinase

A. Structures of the Tyrosinase Active Sites

TYR belong to type 3 copper-containing proteins, which encompass hemocyanins (Hc), catechol oxidase (CO, EC 1.10.3.1), and aureusidin synthase (AUS, EC 1.21.3.6).[9–11] The active site of TYR had been identified very early as a structural analogue[12,13] of that of CO[14,15] and Hc[16] for which 3D X-ray structures were available. It was not until 2006 that Matoba *et al.* confirmed this analogy with the resolution of the 3D structure of the recombinant TYR from *Streptomyces castaneoglobisporus* bacteria (*Sc*TYR).[17] In type 3 copper-containing proteins, the common feature is an active site composed of two copper ions at a distance of about 3.6 Å both coordinated by three *N*-imidazoles of histidine residues.[18] In fungi[19–22] and plant[23] TYRs known to date (Table 1), one of the six histidine residues forms an unusual thioether bond with an adjacent cysteine residue (Figure 3a). The role of this post-translational modification that does not significantly modify the structure of the active site is not clear. Several hypotheses have been made such as the stabilization of the binuclear copper site to facilitate the electron transfer during the catalytic reactions[24] or a role in the copper incorporation process into the *apo*-TYR.[19] However, despite this post-translational modification, the active sites of all TYRs are similar and nearly superimposable (Figure 3b).

No structure of human TYR is yet available with a good resolution.[28] However, two homology models based on *Sc*TYR and *Ipomoea batatas*

Table 1.   TYR 3D structures known to date.

TYRs	Organisms	PDB ID	Cys-His	Ref.
*Sc*TYR	*Streptomyces castaneoglobisporus*	1wx2	No	17
*Bm*TYR	*Bacillus megaterium*	3nm8	No	25
*Ao*TYR	*Aspergillus oryzae*	3w6w	Yes	19
*Ab*TYR	*Agaricus bisporus*	2y9w	Yes	20
*Ms*TYR	*Manduca sexta*	3hhs	No	26
*Mj*TYR	*Marsupenaeus japonicas*	3wky	No	27
*Jr*TYR	*Juglans regia*	5ce9	Yes	23

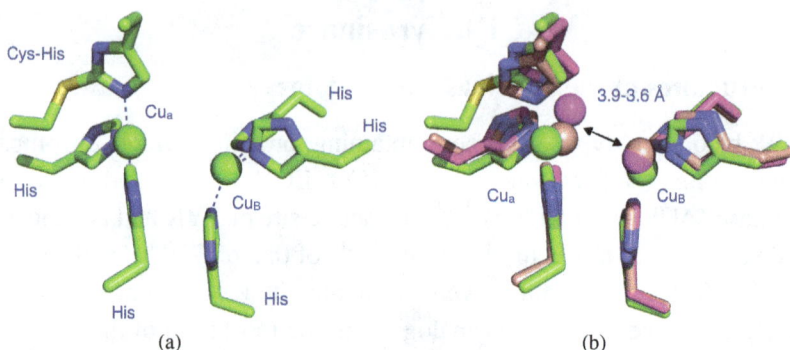

Figure 3. (a) Post-translational modification involving cross-linking of a His ligand with a Cys residue in *Ab*TYR and (b) Active site alignments for *Sc*TYR (salmon), *Bm*TYR (magenta), and *Ab*TYR (green).

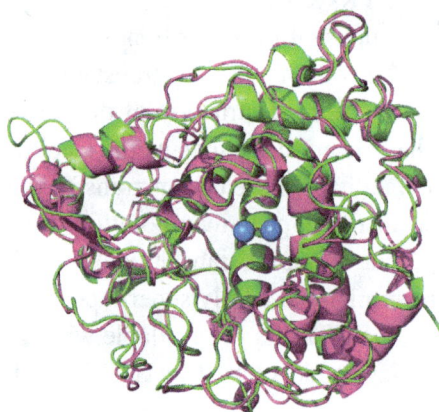

Figure 4. Overlay of the homology model of *Hs*TYR (green) based on TRP1[30] and the one obtains from AlphaFold AI approach (AF-P14679-F1 model_v2) (pink). Copper ions are in blue.

catechol oxidase (*Ib*CO) crystal structures,[29] and on the TYR-related protein (TRP1) crystal structure (*vide infra*)[30] have been generated. More recently, with Alphafold AI interface,[31] a model was also generated that overlaps well with the one based on TRP1 (Figure 4).

The type 3 copper-containing proteins have been shaped by evolution to bind and activate dioxygen: (*i*) O_2 transporters in mollusks and arthropods (Hc) where the bound dioxygen is shielded by the protein matrix and

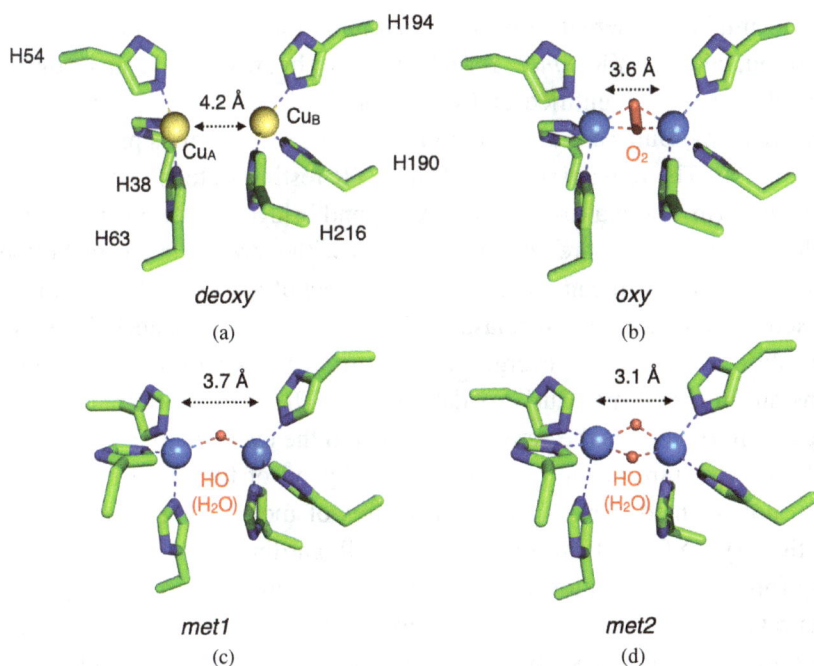

Figure 5. Structures of the TYR active site from *Streptomyces castaneoglobisporus* bacteria (*Sc*TYR) showing: (a) the dicopper(I) *deoxy* form (PDB ID: 2ahl), (b) the dicopper(II) *oxy* form (PDB ID: 1wx4), (c) the dicopper(II) *met1* form (PDB ID: 2zmx), and (d) the dicopper(II) *met2* form (PDB ID: 2zmy) that are involved in the catalysis. Cu(I) are displayed as yellow balls and Cu(II) are displayed as blue balls.

therefore not reactive, (*ii*) mono-oxygenases in *ortho*-hydroxylation of phenols (TYR and AUS) and (*iii*) di-oxygenases in catechol oxidation (TYR and CO) in which activated dioxygen can react with the substrate. From all the collected spectroscopic data and 3D structures from diverse TYR sources, three different forms have been identified, the *met*, *oxy*, and *deoxy* forms (Figure 5).

B. Catalytic Mechanism of Tyrosinase

The TYR resting state is composed by 85% of the *met1* form, in which a hydroxyl ligand holds the two Cu(II) ions approximately 3.2–4.0 Å apart. Recent QM/MM calculations seem to suggest that an aqua (H_2O) bridging ligand might be energetically more favorable than a hydroxido ion.[32] In addition, with ScTYR, a *met2* form having two hydroxido or aqua bridging ligands was also identified and characterized.[33] From X-ray absorption spectroscopy, Bubacco *et al.*[34] reported that the *met2* form is predominant in solution. The remaining 15% of the TYR resting state is composed by the *oxy* form where a $(\mu\text{-}\eta^2\text{:}\eta^2)$-peroxo ligand bridges the two Cu(II) ions. While the *met* forms are only able to react with catechol in a catecholase cycle, the *oxy* form can react both with catechol and phenol in an intermixed catecholase and cresolase cycles (Figure 6). The catecholase and cresolase cycles both converge to the *deoxy* form where the two copper ions are in the formal Cu(I) oxidation state. The *deoxy* form reacts with dioxygen to produce the active *oxy* form and the cycle continues.

The structure of the $(\mu\text{-}\eta^2\text{:}\eta^2)$-peroxo ligand of the *oxy*-TYR is now well documented. Several X-ray structures of model compound[35,36] and of the *oxy*-TYR confirm this structure.[17] Regarding the reactivity of the *oxy*-form, Karlin's pioneering work on chemical model complexes has revealed the electrophilic behavior of the $(\mu\text{-}\eta^2\text{:}\eta^2)$-peroxo ligand.[37] The same feature was observed also with the *oxy*-TYR. Thanks to a

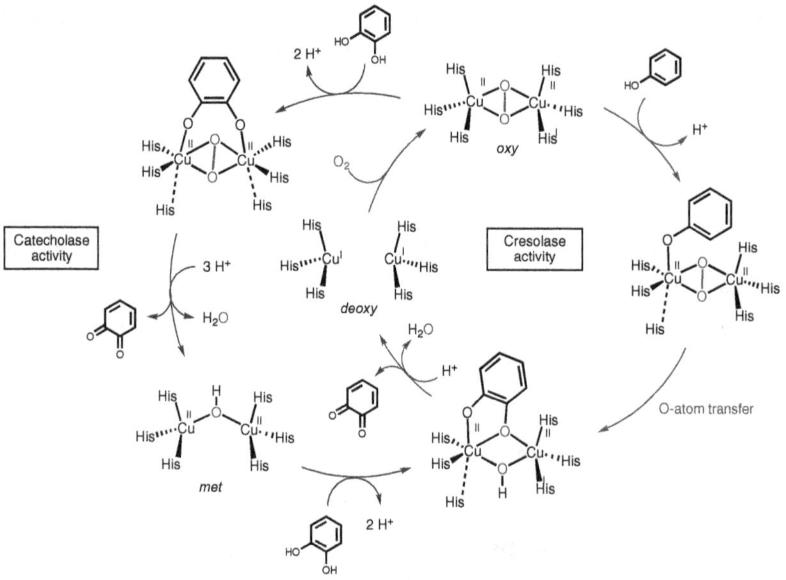

Figure 6.   General mechanism for the cresolase and catecholase TYR activities.

kinetic study of the TYR catalyzed phenol hydroxylation, Itoh *et al.* have shown that the rate-determining step is the O-atom transfer occurring through an electrophilic aromatic substitution mechanism.[38]

With data from different groups, the TYR mechanism is quite well established[39] (Figure 7). The first step is the formation of a phenolate bond on the Cu_A site of the *oxy*-TYR. Two mechanisms have been proposed for the formation of the phenolate bond on the Cu_A site. The first was proposed by Decker and Tuczek on the base of crystallographic data of *Sc*TYR.[40] The authors proposed a pre-orientation of the phenolate by π-staking interaction with the H194 of the Cu_B site followed by a phenolate sliding toward Cu_A (Figure 7A). With *Ab*TYR, a similar proposal involving a pre-orientation of the phenolate by a π-staking interaction with a histidine residue of the Cu_B site was proposed by Dijkstra.[41] Based on crystallographic data on *Ao*TYR, Itoh *et al.* proposed another mechanism for the formation of the Cu_A-phenolate. While Decker and Tuczek propose a shift from phenolate toward Cu_A, Itoh proposes an opposite movement, that of Cu_A toward phenolate located in a stabilizing enzymatic pocket[42] (Figure 7B). This latter proposal is supported by several strong evidences obtained with *Ao*TYR mutants, copper-depleted or zinc-replaced *Ao*TYR.

Both mechanisms lead to phenolate binding to Cu_A site in *trans* position relative to H63 for *Sc*TYR with a concomitant elongation of the H63-Cu_A distance (H103 for *Ao*TYR). Phenolate binding impulses a 90° rotation of the O–O plan of the peroxo group toward the aromatic ring that allows the peroxide σ^* orbital to overlap with the π orbital of the phenolate *ortho*-position thus facilitating the electrophilic attack of the peroxide by the phenolate.[39,40] This process leads to the formation of a catecholate bridging the two copper centers. The structure of this catecholate that is not well defined could be a κ^2-catecholate on Cu_A (**a**), or a μ-1,4-catecholate (**c**) or the combination of both, that is ($\eta^2 : \eta^1$)-catecholate (**b**) (Figure 7C).

C. Tyrosinase-Related Proteins TRP1 and TRP2

Two other enzymes, TRP1 and TRP2, closely related to TYR, are involved in the biosynthesis of eumelanin pigments.[30] TRP1 and TRP2 share ~70% amino acid homology, suggesting that they originate from a

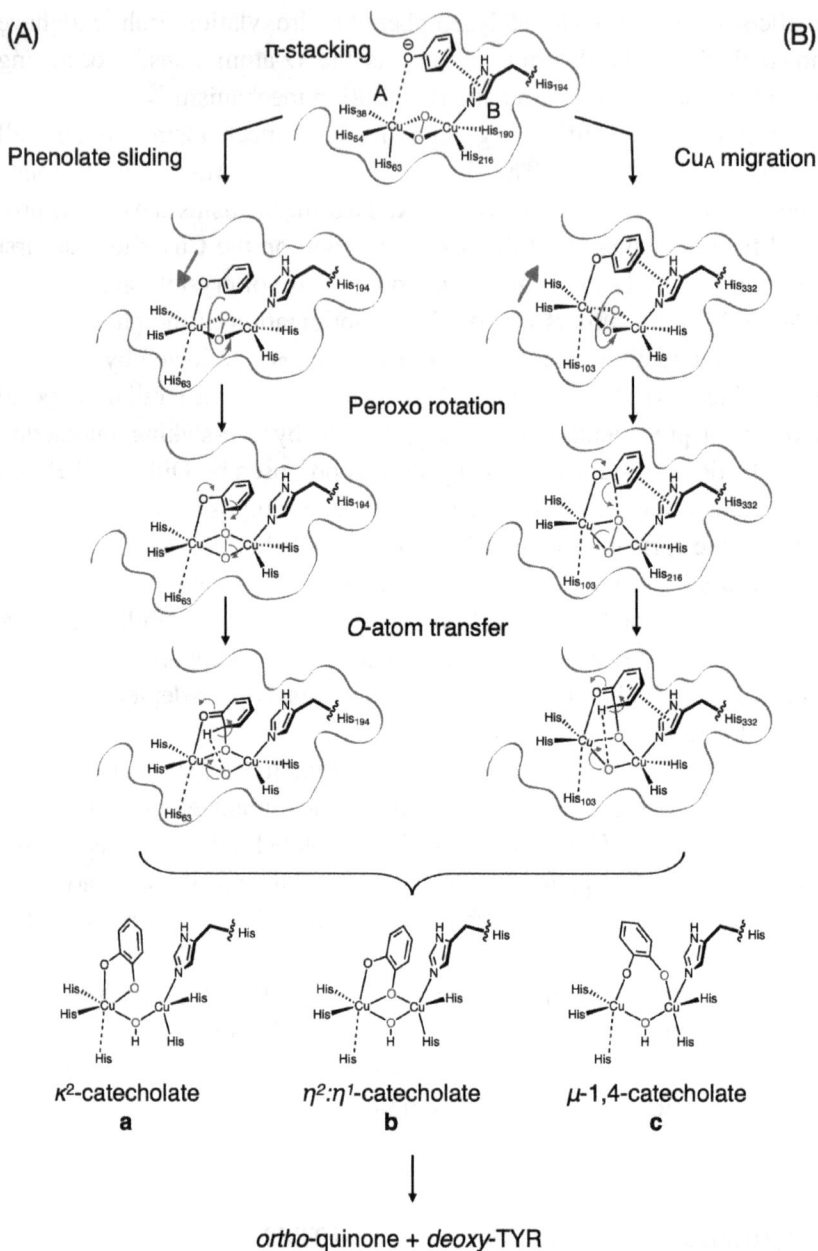

Figure 7. Mechanisms proposed by (A) Decker and Tuczek[39] on *Sc*TYR and (B) Itoh on *Ao*TYR[42] with putative structures (a)–(c) for the catecholate-TYR form.

common ancestral gene. While TRP2 was very early identified as zinc-containing analogue involved in DHIC formation, TRP1 was firstly identified as copper-containing analog having a catecholase activity and involved in the oxidation of DHI and DHICA into the eumelanin pigments precursors.[43] However, the recent 3D structure of TRP1 solved by Soler-López and Dijkstra et al.[44] revealed, against all expectations, the presence of Zn(II) instead of Cu(II) ions in the active site thus ruling out the possible role of TRP1 in DHICA oxidation. Since this publication, TRP1 function has become enigmatic, and several hypotheses have been formulated such as a role in stabilization of *Hs*TYR as proposed by Dolinska et al.[45] or its contribution in increasing the ratio between eumelanin and pheomelanin by promoting the synthesis of eumelanin as proposed by Ishikawa et al.[46] But an unresolved question still remains: "could TRP1 integrate Cu(II) ions *in vivo*?" Indeed, *ex vivo* experiments have shown that catecholase activity (DHIC oxidation) can be partially conferred on the human TRP1 when its Zn(II) ions are replaced by Cu(II) ions.[44] Another possibility is that like TRP2 (Figure 8b), TRP1 may have an

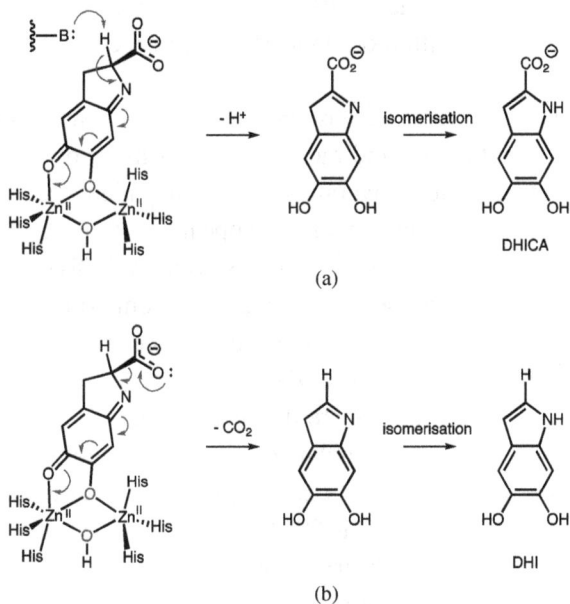

Figure 8. Reaction catalyzed by TRP2 (a) and reaction which may be catalyzed by TRP1 (b).

activity linked to the Lewis acid property of the Zn(II) ions such as the decarboxylation of Dopachrome into DHI that is supposed to be non-catalyzed (Figure 8b).

IV. Dysfunction in Tyrosinase Activity

A. Pathologies Linked to TYR Dysfunction

Being involved in the biosynthesis of melanin, TYR is associated with cutaneous pigment disorders essentially linked to (*i*) the hypopigmentation of the skin such as in albinism and vitiligo or (*ii*) the hyperpigmentation of the skin such as in melanoma, melasma, and solar lentigo.[47]

Overexpressed during tumorigenesis,[48] TYR has been demonstrated to be a sensitive marker for melanoma.[49] Melanoma, which is the most serious type of skin cancer, is a malignant tumor arising from melanocytes. During the tumorigenesis process, melanocytes lose their interactions with epidermis cells (keratinocytes) that result in uncontrolled proliferation followed by an invasion of malignant cells. Melanoma is the deadliest form of skin cancer (20–25% mortality), ranked 17th among cancers of all genders with more than 150,000 new cases of melanoma in 2020.[50]

It was observed that an increase in melanogenesis may affect immune responses to chemotherapy and radiotherapy for melanoma. A high level in melanogenesis shortens overall survival in patients with metastatic melanoma. Inhibition of melanogenesis appears a rational approach to the therapy of metastatic melanoma.[51] In that sense, TYR DNA vaccines have been widely developed for the treatment of melanoma since they are much more stable than conventional vaccines and since they have less severe side effects.[52] Treatments by TYR DNA vaccines are generally associated with adjuvants to induce the innate immune system. Because of its over-expression mainly confined in melanocytes during tumorigenesis, TYR is a potential molecular target for the development of specific inhibitors that could be used as adjuvant to TYR DNA vaccine in melanoma therapy.[53,54] This is supported by recent observations suggesting that melanin secreted from melanoma cells would be responsible for the immunomodulation of tumor microenvironment. An increase of the level of secreted melanin would promote tumor angiogenesis, thus sustaining melanoma progression and metastasis.[55,56] Since the reduction/suppression of melanogenesis is supposed to restore the sensitivity of cancer cells to immuno-, radio-, or chemotherapies, the use of selective inhibitors of TYR and TRP1,2 as adjuvants represents a realistic strategy for melanoma therapy.[57–59]

B. Skin Whitening

Depigmenting agents are commonly used in dermatology in the treatment of hyperpigmentary disorders.[60] Kojic acid in association with hydroquinone is prescribed in Europe under medical supervision, in the treatment of post-inflammatory hyperpigmentation[61] and in melasma.[62] Unfortunately, the effectiveness of these agents has led to their use being diverted, toward non-therapeutic purposes, in voluntary depigmentation with the objective of lightening the natural complexion of the skin.[63,64] This practice, which is widespread in sub-Saharan Africa, has become a real public health problem as it leads to serious dermatological and systemic complications. The ingredients historically encountered in cosmetic products intended to lighten the skin are hydroquinone, mercury derivatives, and corticosteroids. Recent studies have demonstrated the harmful effects of these active ingredients for human health with the development of ochronosis, skin complications, addiction to steroids, periorbital hyperchromia, facial lesions, or even kidney failure.[65,66] Although better controlled, this practice is also rampant in Asia (China, Japan, and India) where it is estimated that these products represent nearly 10% of the cosmetics market.

In this context, measures must be taken to curb this tendency. Facing the difficulty of fighting these dangerous practices through laws, the development of harmless and inexpensive bleaching agents accessible to all could be a solution to this problem.

V. Tyrosinase Inhibition

Because inhibition of TYR is a well-established approach to control *in vivo* melanin production, the development of inhibitors has a huge economic and industrial impact in medicine and cosmetics (*vide supra*) but also in agriculture to reduce the browning of fruits and vegetables.[67] This impact is assessed by the increasing number of review articles on TYR inhibitors reported in the recent literature.[68–74] A large part of this literature is devoted to naturally occurring compounds such as flavonoids and polyphenolic compounds,[71] triterpenes, alkaloids, as well as mixture extracts from plants[73–75] and peptides.[72]

A. Variation in TYR Sources

Due to the commercial availability of the fungal *Ab*TYR, most TYR inhibition studies have been conducted on this TYR as a model for the human form (*Hs*TYR). The molecules resulting from these tests are often incorrectly designated as potential TYR inhibitors that can be used in human pathologies.[76] Although the structure of the first coordination sphere of the dinuclear active site is conserved in all the TYRs, they present significant structural and cellular localization differences.[11] In addition, although there are very strong intra-species sequence homologies, the inter-species differences are important. For example, within a region of 48–49% sequence coverage, fungal TYRs have only 22–24% identity with mammalian ones. It should be mentioned that the fungal enzyme *Ab*TYR is a cytosol-soluble heterodimer whereas *Hs*TYR is a 529–amino acid glycoprotein anchored to the membranes of melanosome vesicles by a 21–amino acid transmembrane domain. Some differences reside also at the level of the second coordination sphere of the metal center with residues of polar or hydrophobic character controlling the accessibility to the active site and substrate's selectivity (activity controllers).[77] As it can be seen in Figure 9, recent 3D structures highlight the huge differences in active site accessibility between *Ab*TYR, *Sc*TYR, and *Hs*TYR.

B. Transition-State Analogue (TSA) Inhibitors

One way to selectively inhibit TYR activity is to target its copper binuclear active site that is unique within human metalloenzymes. Some of TYR inhibitors such as phenylthiourea (**PTU**),[78] thiosemicarbazones (**TSC**),[79–89] L-mimosine (**Mim**),[90,91] Kojic acid (**KA**),[92,93] tropolone (**Trop**)[94] and thujaplicin (**Thuj**),[95,96] 2-hydroxypyridine *N*-oxide (**HOPNO**),[97] and

Top view   Longitudinal sections

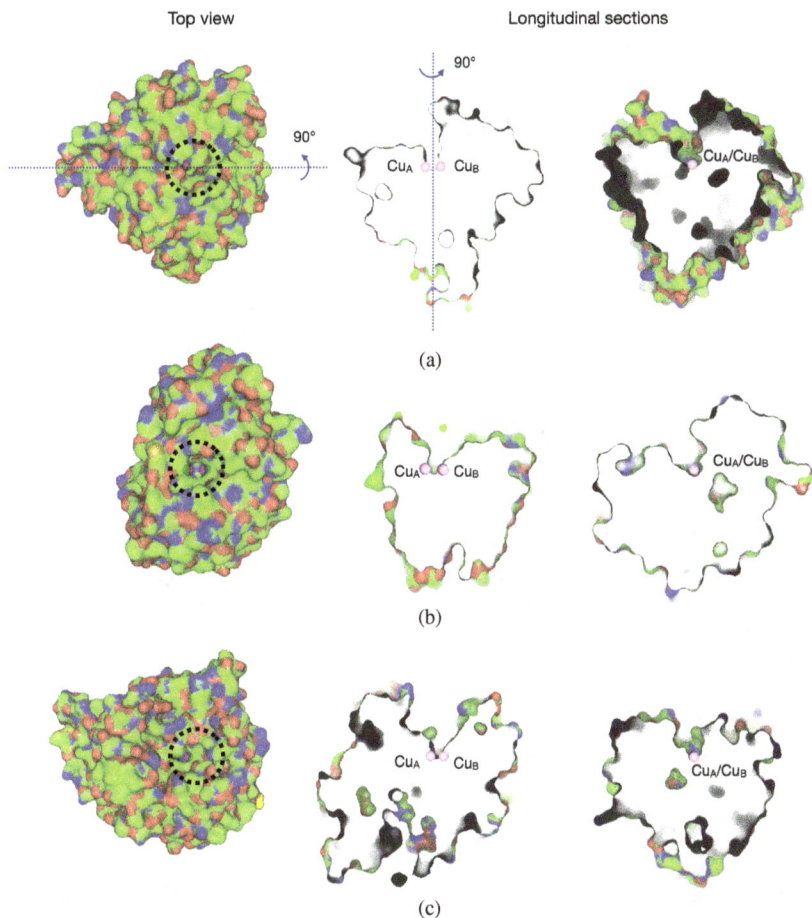

(a)

(b)

(c)

Figure 9. Accessible surface of TYR structures with longitudinal sections showing the accessibility of the copper active site; (a) AbTYR (PDB ID 2y9x), (b) ScTYR (PDB ID 2ahl), and (c) HsTYR (Homology model).[30]

2-pyridinethiol *N*-oxide (**HSPNO**)[98] fulfil this objective. These inhibitors are chelators that all target the dinuclear copper site of TYR (Figure 10). Among them, kojic acid, L-mimosine, tropolone, thujaplicin, **HOPNO**, and **HSPNO** are also TSAs of the *O*-atom transfer reaction. Indeed, they present structural analogies with catechol, but they are not oxidizable (Figure 10b). These molecules present a double interest, they are both capable of modulating TYR activity (medicinal interest) and they are likely to

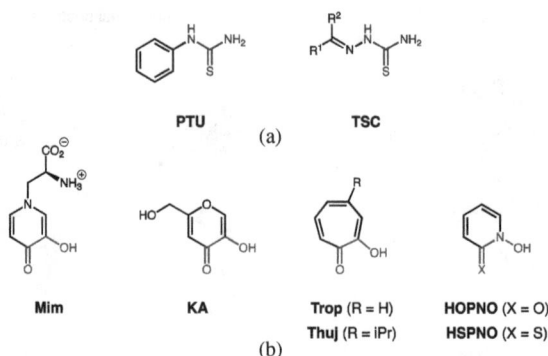

Figure 10. TYR Inhibitors that target the copper dinuclear active site: (a) metal chelating by the thiourea moieties in blue and (b) TSAs.

Table 2. Inhibition constant (K_I) for selected TSA compounds.

	Fungus	Bacterial		Mammalian
Inhibitors	*Ab*TYR	*Sa*TYR	*Bm*TYR	*Hs*TYR
Mim	C, 8 μM[99]	C, 54 μM[100]	nd	C, 10.3 μM[100]
KA	C, 4.3 μM[101]	C, 109 μM[100]	M, 3.5 and 150 μM[102]	M, 350 and 730 μM[100]
HOPNO	C, 1.8 μM[97]	C, 7.7 μM[103]	nd	C, 128 μM[104]

C, competitive; M, mixed; nd, not determined.

provide information on the structure of the catecholate-TYR intermediate (structural and mechanistic interest).

Kojic acid, L-mimosine, and **HOPNO** inhibitors for which K_I constants are available on three sources of TYR, fungus (*Ab*TYR), bacterial (*Sa*TYR and *Bm*TYR), and human (*Hs*TYR) have been used as reference to better understand the mechanism of TYR inhibition by TSA compounds. As expected, the selected TSA compounds behave as competitive inhibitors for the substrate L-DOPA (Table 2). Kojic acid behaves as mixed inhibitor for the *Hs*TYR and *Bm*TYR but with a competitive component much lower than the uncompetitive one. The inhibitory properties estimated from the K_I reveal significant differences between the three sources of TYR, and the best inhibitions are always obtained with the fungal *Ab*TYR. However, L-mimosine, which is the best inhibitor for the

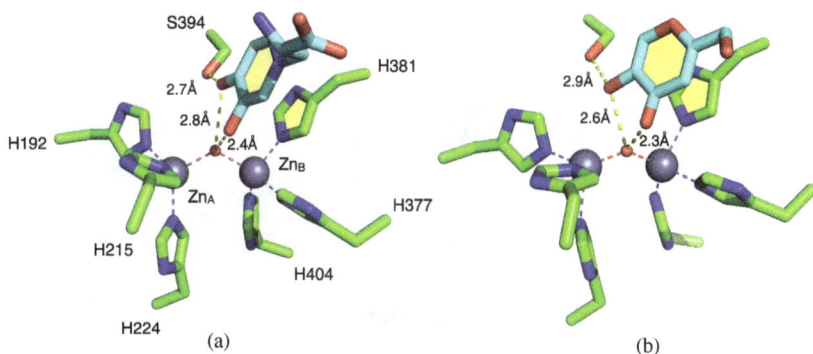

Figure 11. TRP1 active site with (a) L-mimosine (PDB ID: 5m8n) and (b) kojic acid (PDB ID: 5m8m).[44] The coloring of the cycles in yellow shows the stabilizing π-stacking interactions.

*Hs*TYR, shows comparable inhibitory properties for the fungal and the human TYRs.

1. *L-Mimosine*

A TRP1 structure with L-mimosine as ligand has been recently solved by Soler-López and Dijkstra *et al.*[44] Surprisingly, this structure reveals that L-mimosine does not directly interact with the Zn(II) ions but that the two *O*-atoms of L-mimosine are H-bonded to the water molecule bridging the two Zn(II) ions (Figure 11a). The structure also points out additional interactions such as π-stacking with the His381 and H-bonding of one of the *O*-atom with Ser394.

2. *Kojic acid and derivatives*

Based on ESEEM (electron spin echo envelope modulation),[105] HYSCORE (Hyperfine sublevel correlation),[106] and XAS (X-ray absorption spectroscopy)[34] coupled to MXAN simulations,[107] all performed on *Sa*TYR, Bubacco *et al.* proposed a model where kojic acid is binding in a $\eta^2{:}\eta^1$-KA coordination mode, that is, as bidentate ligand to Cu_B with one *O*-atom axially coordinated and the second bridging the two copper ions in an equatorial coordination position in place of one of the two water (OH) bridging *O*-atoms (Figure 12a).

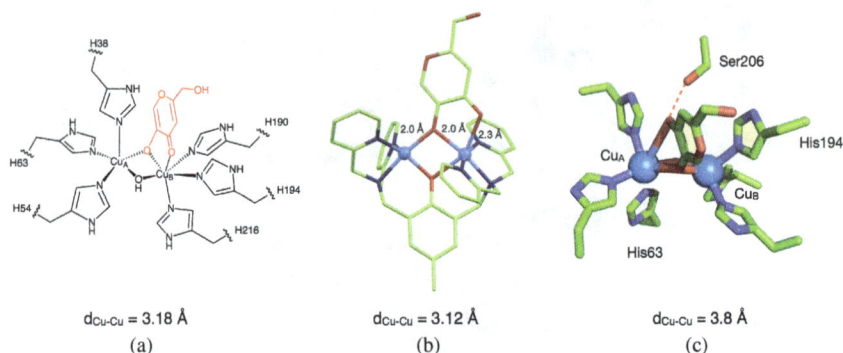

dCu-Cu = 3.18 Å	dCu-Cu = 3.12 Å	dCu-Cu = 3.8 Å
(a)	(b)	(c)

Figure 12. Kojic acid bound to a binuclear copper center, (a) SaTYR-KA adduct according to X-ray absorption spectroscopy studies[34] (ScTYR numbering); (b) X-ray crystal structure with a copper model complex,[108] and (c) QM/MM dynamics simulations for ScTYR-KA adduct.[108] Stable position is obtained after 2.5 ps of QM/MM dynamics simulations. The coloring of the cycles in yellow shows the stabilizing π-stacking interaction and the dotted red line the H-bond interaction between S206 and KA.

The $\eta^2{:}\eta^1$-KA coordination mode has been also reported in model chemistry (*vide infra*) by Belle *et al.*[108] With a binuclear Cu(II) complex[109] known to be a structural and functional model of TYR diphenolase activity, an adduct with kojic acid that exhibits this binding mode was obtained and characterized by X-ray diffraction (Figure 12b).

While the XAS studies identify a $\eta^2{:}\eta^1$-KA coordination mode for kojic acid,[34] it is not obvious on which Cu-ion (Cu_A *vs.* Cu_B) kojic acid is bidentatety bound. QM/MM simulations performed by Jamet *et al.*[108] have tried to clarify the situation. Since no X-ray structure of the SaTYR was available, QM/MM simulations were performed using the structure of ScTYR that shares more than 82% sequence identity with SaTYR. Therefore, both possibilities were tested in dynamic QM/MM simulations. Kojic acid as bidentate ligand to Cu_A was found to be more stable than as bidentate ligand to Cu_B. Moreover, the simulations highlight π-stacking interaction with the His388 and H-bonding interaction of one of the *O*-atoms with Ser206, interactions already observed with L-mimosine in TRP1 (*vide supra*). During the simulation with kojic acid as bidentate ligand to Cu_A, the His63 coordinated to Cu_A, moves to a distance around 3.76 Å from the copper leading to a penta-coordinated

Cu_A-ion (Figure 12c). This agrees with EXAFS (Extended X-Ray Absorption Fine Structure) data,[34] that support histidine decoordination in SaTYR-**KA** complex. During the simulation with kojic acid as Cu_B bidentate ligand, the Cu_B–O^3 distance increases until decoordination and the O^3-atom shift toward Cu_A to become a Cu_A-bidentate ligand.

More recently 3D X-ray structure of the *Bm*TYR in complex with kojic acid was reported by Fishman *et al.*[25] Surprisingly, in this structure, kojic acid is oriented with its hydroxymethyl group in the direction of the active site at a relatively remote distance of 7 Å, which is the opposite of what was reasonably expected and observed with *Sa*TYR (Figure 13b). Kojic acid is stabilized by interactions with Phe197, Pro201, Asn205, and Arg209. The same group reported later a novel structure where kojic acid is oriented as expected toward the active site (Figure 13a). But in this structure, kojic acid does not directly interact with the Cu(II) ions[102] (Figure 13a). The hydroxyl and carbonyl groups are found at distances greater than 3.3 Å from the Cu_A center, which does not allow to consider coordination bonds. However, kojic acid is stabilized by π-stacking interaction with His208 and H-bonding interaction with the water molecule bridging the two Cu(II) ions, a situation already encountered for L-mimosine in interaction with TRP1. Furthermore, Soler-López and

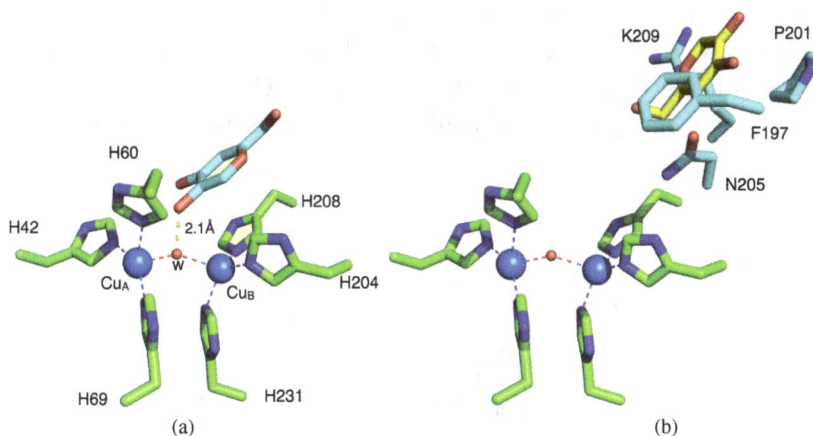

Figure 13. *Bm*TYR active site with kojic acid (a) oriented to the active site (PDB ID: 5i38)[102] and (b) in entrance to the active site (PDB ID: 3nq1).[25] The coloring of the cycles in yellow shows the stabilizing π-stacking interaction.

Dijkstra *et al.* reported a structure of TRP1 with kojic acid in which the same features are retrieved[44] (Figure 11b). These two orientations of kojic acid support the mixed inhibition observed for *Bm*TYR (Table 2). Kojic acid in the active site precludes the substrate's fixation while in the other position, it reduces substrate accessibility and product release.

Apart from the binding mode of kojic acid to the *met*-TYR, these latest results raise fundamental questions about the hydroxylation mechanism catalyzed by TYR, namely: (*i*) if the substrate binds directly to the metal center or if the oxidation takes place without binding and (*ii*) if the substrate binds to the metal center on which site does it bind, to Cu_A or Cu_B? Until now, all the mechanistic discussions agreed on the hypothesis that the substrate had to be fixed to the metal center, by a yet not elucidated mechanism (Figure 7a *vs.* 7b),[39,42] to induce the rotation of the peroxo group, which triggers the reaction.[40]

Crystallization results with inhibitors were achieved by soaking TYR crystals in an inhibitor solution. These conditions are however far from those used during catalysis, which raises questions about the relevance of these structures for understanding the catalytic mechanism of TYR. Nevertheless, they can be seen as inspiration for the development of increasingly effective inhibitors. Starting from the observation of two different binding sites for kojic acid, ditopic inhibitors have been synthesized and studied (**KA-TSA**[110] and **bis-KA**[111]) (Figure 14). These inhibitors were designed to interact with both high and low affinity KA-sites in *Bm*TYR, and probably in *Hs*TYR. Unfortunately, they displayed only slightly improved inhibitory properties as compared to kojic acid.

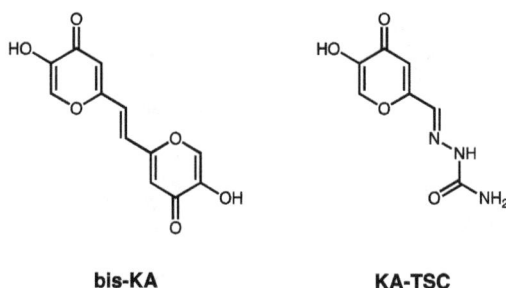

bis-KA   **KA-TSC**

Figure 14.   Ditopic inhibitors based on kojic acid.

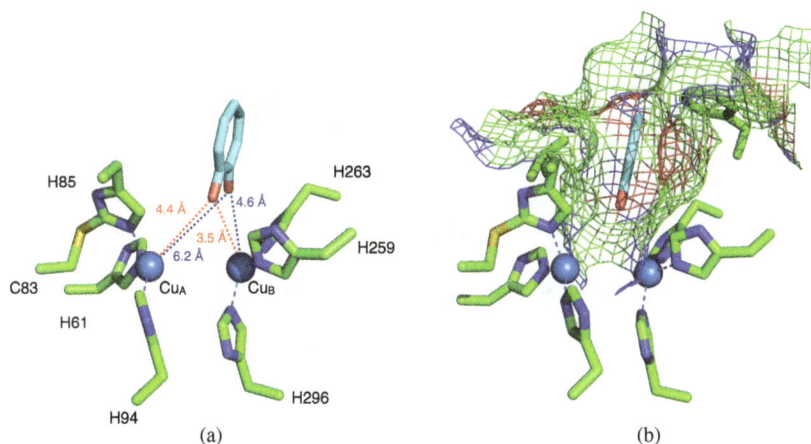

Figure 15. Tropolone in interaction with (a) *Ab*TYR active site (PDB ID: 3y9x)[20] and (b) showing the hydrophobic pocket above the active site where tropolone is stacked.

3. *Tropolone and derivatives*

Tropolone is a slow-binding, reversible *Ab*TYR inhibitor that only inhibits *oxy*-TYR.[112,113] The first crystal structure of *Ab*TYR in interaction with tropolone was reported by Dijkstra *et al.*[20] In this structure, tropolone binds to a large cavity (Figure 15b) without the need of conformational change of the protein to allow inhibitor's entrance. Since the *O*-atoms of tropolone are quite away from the copper center (>3.5 Å, Figure 15a), it was concluded that this tropolone-binding mode represents a pre-Michaelis complex that foreruns dioxygen fixation.

TRP1 also binds tropolone in a κ^2-coordination mode on the Zn_A ion as shown by crystal structures.[44] Like other TSA inhibitors, tropolone binding occurs by aromatic π-stacking interactions with H381, ligation of the keto- and hydroxy groups to the Zn_A ion, and H-bonding interaction with S394 (Figure 16).

α, β, and γ-thujlaplicins, naturally occurring compounds containing a tropolone framework, exhibit interesting inhibition properties of *Ab*TYR and *Hs*TYR (Figure 17).[114,115] In the case of *Ab*TYR, with the exception of α-thujlaplicin, which isopropyl group precludes correct positioning in the active site, both β and γ-thujlaplicins exhibit IC_{50} in the nanomolar range. They are therefore 700-fold more active than the kojic acid

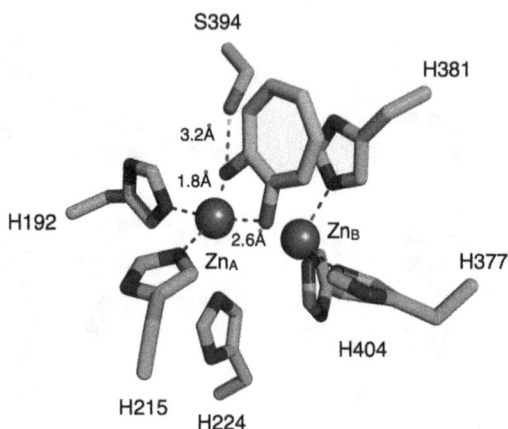

Figure 16. Tropolone in interaction with the TRP1 active site (PDB ID: 5m80).[44] The coloring of the cycles in yellow shows the stabilizing π-stacking interaction.

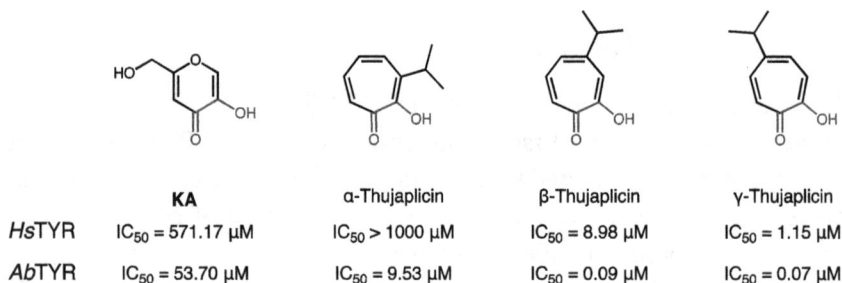

	KA	α-Thujaplicin	β-Thujaplicin	γ-Thujaplicin
*Hs*TYR	$IC_{50} = 571.17$ μM	$IC_{50} > 1000$ μM	$IC_{50} = 8.98$ μM	$IC_{50} = 1.15$ μM
*Ab*TYR	$IC_{50} = 53.70$ μM	$IC_{50} = 9.53$ μM	$IC_{50} = 0.09$ μM	$IC_{50} = 0.07$ μM

Figure 17. Thujlapines as TYR-inhibitors.

reference. For *Hs*TYR, the tendency is the same; γ-thujlaplicin exhibits IC_{50} in micromolar range, 500-fold more active than the kojic acid reference.

4. *HOPNO inhibitor*

In the absence of 3D structures of TYR with the **HOPNO**, XAS spectra were recorded on *Sa*TYR in complex with **HOPNO** (Figure 18). Significant modifications were observed in the XANES part of the XAS spectra upon addition of **HOPNO** to the *Sa*TYR indicating an interaction of **HOPNO** involving a rearrangement around the coordination sphere of

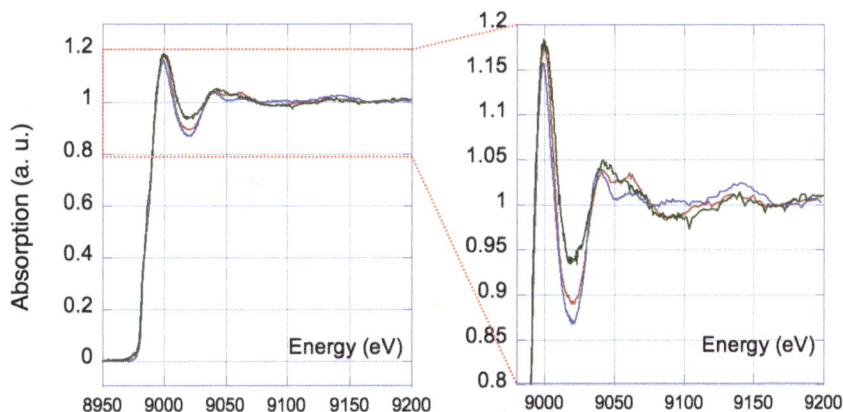

Figure 18.   XAS of SaTYR alone (green) and with **KA** (blue) and **HOPNO** (red).

the Cu(II) ions. The same behavior was also observed upon addition of kojic acid, but significant differences indicate that **OPNO**[a] and kojic acid bind differently the copper center.

The biomimetic approach in developing small molecular complexes reproducing the main structural characteristics of the active site and the reactivity of metalloenzymes allows to better understand their structure–function relationships [Chapters 3 and 4].[116] With this strategy, Belle *et al.* has reported the synthesis and characterization of a copper complex based on H-BPMP ligand that reproduces: (*i*) the copper coordination with 3 *N*-atoms and a μ-phenolate and μ-hydroxy groups allowing a copper–copper distance of 2.966 Å and (*ii*) the catecholase activity of TYR and CO[109] (Figure 19).

On the contrary of the kojic acid adduct to BPMP(Cu$_2$)OH complex that features a η^2:η^1-KA coordination mode[107] (*vide supra*), it was observed a **HOPNO** adduct to BPMP(Cu$_2$)OH complex featuring a κ^2-OPNO coordination mode on only one of two copper ions.[97] This highlights the differences between inhibitors interaction at a dinuclear active site and the difficulty to draw general conclusions for all TSA inhibitors.

---

[a] Since the pK_a for the N–OH group of **HOPNO** is 6.07 [98], in the conditions used (pH = 7) **HOPNO** is mainly in the deprotonated form **OPNO**.

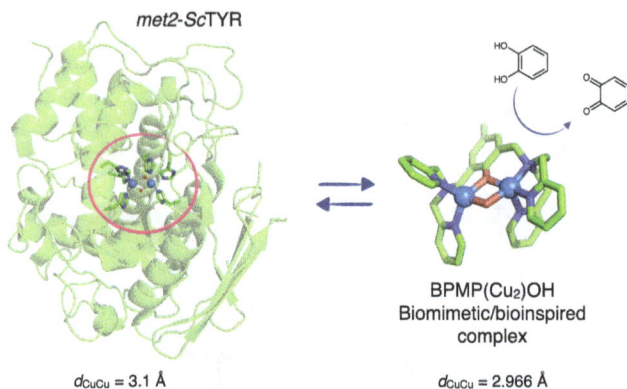

Figure 19.   Biomimetic strategy in TYR study.

An important factor in the formation of TSA adducts is the geometry around the Cu(II) ions. Although the coordination around the two Cu(II) ions is the same in all TYR active sites, there are significant differences in their accessibility but also by their geometry, which reflects their reactivity. The geometry of penta-coordinated complexes, that is, square-based pyramid (SP) *versus* trigonal bipyramid (TBP) can be characterized by the τ factor, which is determined by taking the difference of the two largest angles in the coordination sphere of the metal divided by 60. A $\tau = 1$ is characteristic of a pure TBP, whereas a $\tau = 0$ is of a pure SP geometry.[117] In the *met2*-form of *Sc*TYR, the two Cu(II) ions present distorted SP geometries as reflected by their τ factors (Cu$_A$, $\tau = 0.3$ *vs.* Cu$_B$, $\tau = 0.5$, Figure 20a). The distortion is more pronounced for Cu$_B$ than is between SP and TBP geometries.

A similar distortion is also observed in the model complex BPMP(Cu$_2$) OH, in which one Cu(II) ion is in a pure TBP geometry while the other exhibits a distorted SP geometry (Figure 20b). The distortion in coordination can be controlled by the size of the metallacycle of the chelate ligands around the Cu(II) ions. The size of the metallacycle depends on the number of CH$_2$ group (methano *vs.* ethano bridge) between the *N*-tertiary amino group and the pyridine nucleus. The dinuclear complex BPEP(Cu$_2$) OH featuring four 6-membered ring metallacycles, has one Cu(II) ion in a pure SP geometry ($\tau = 0$) while the other is in distorted SP geometry ($\tau = 0.3$) (Figure 20c). The **HOPNO** adduct features a μ-1,4-OPNO

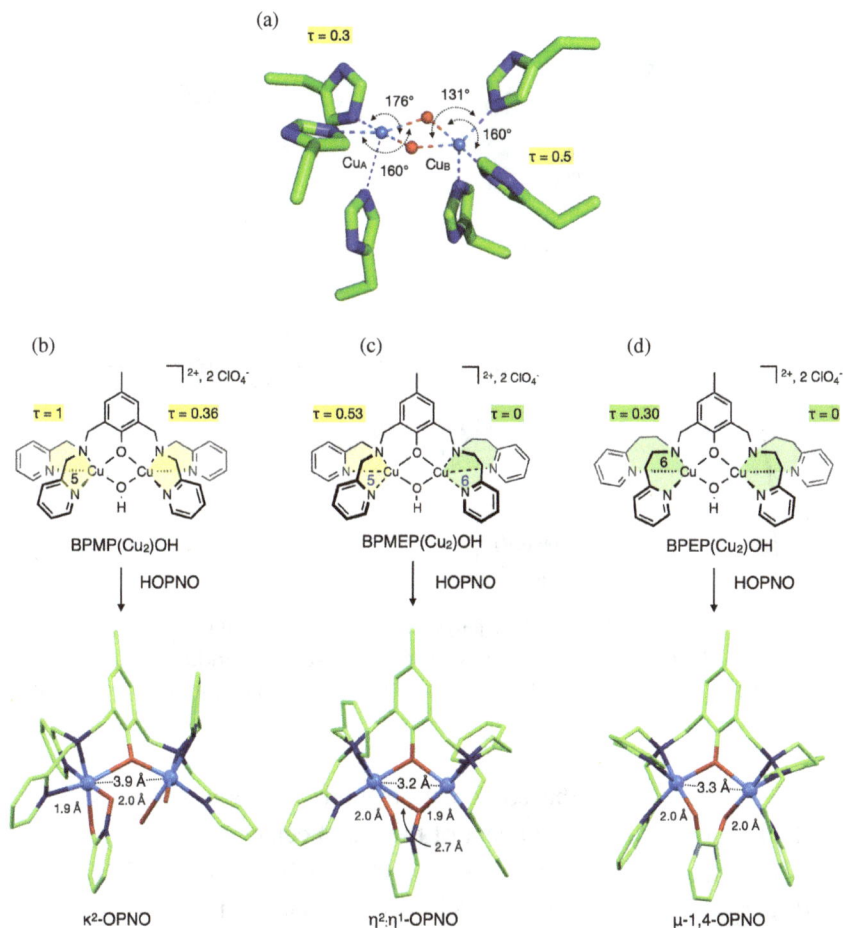

Figure 20. OPNO ligand binding modes on dicopper(II) center. (a) met2-ScTYR active site (PDB ID: 2zmy[17]) with the τ factor[117] for the two copper centers; (b,c) the three different coordination mode obtained with model complexes.

coordination mode.[98] The dinuclear complex BPMEP(Cu$_2$)OH featuring two 6- and two 5-membered ring metallacycles, has one Cu(II) ion in pure SP geometry ($\tau = 0$) while the other exhibits a distorted SP geometry ($\tau = 0.53$) (Figure 20d). Its **HOPNO** adduct features a third coordination mode, that is, η^2:η^1-OPNO.[103] With three slightly different ligands that impose a certain coordination geometry around the Cu(II) ions, it is possible to control the coordination mode of the **HOPNO** inhibitor. These

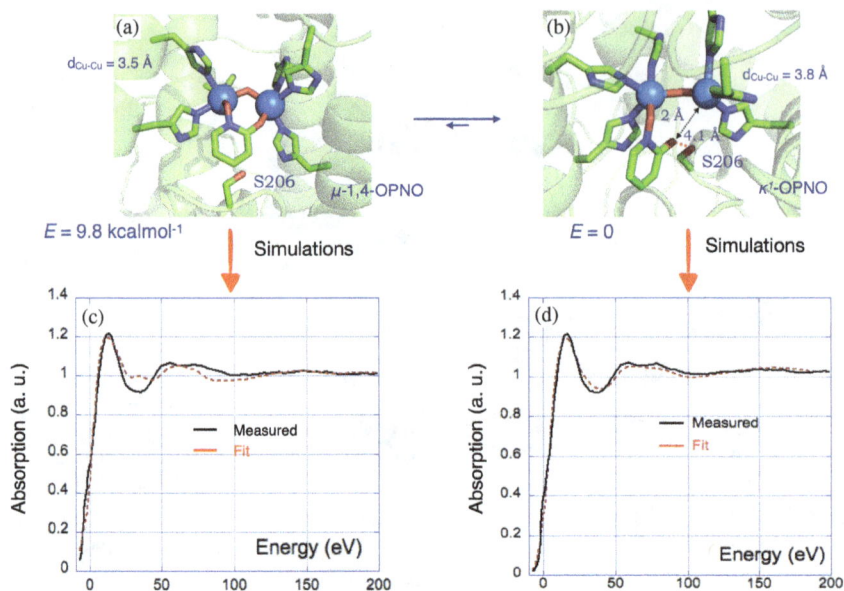

Figure 21. The two minimized structures (a) the μ-1,4-**OPNO** adduct and (b) the κ^1-**OPNO** adduct. (b) XAS simulation of the μ-1,4-**OPNO** adduct and (d) of the κ^1-**OPNO** adduct. XAS of *Sa*TYR (black plain line) and simulation (red dotted line).

results emphasize that the coordination mode of **OPNO** is sensitive to small changes in the environment of the metal center.

To decipher the binding mode of **OPNO** ligand onto TYR, Jamet *et al.* performed QM/MM calculations using data from the model complexes exhibiting the three different **OPNO** binding modes. Starting from these binding modes, the systems converge to two structures, the μ-1,4-**OPNO** and at 10 kcal/mol lower the one with **OPNO** as monodentate ligand that establishes a H-bond with a S206 residue (Figure 21a and b). From these structures, the simulation of the XANES spectra have been performed using MXAN software.[107] The simulation for the μ-1,4-**OPNO** adduct led to a plot (Figure 21c, red dotted line) that does not overlap correctly with the measured XAS of the *Sa*TYR-**OPNO** adduct (Figure 21c, black plain line). Better results were obtained with the simulation of the κ^1-**OPNO** adduct. In that case the simulation better reproduces the measured XAS of the *Sa*TYR-**OPNO** adduct (Figure 21d red

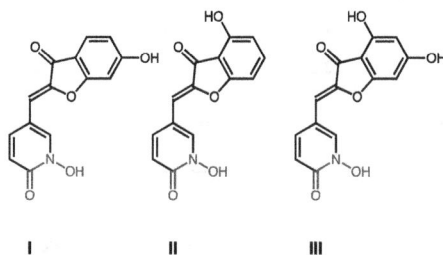

Figure 22.   **HOPNO**-embedded aurones used as TYR inhibitors.

dotted line *vs.* black plain line). These results tend to confirm a dynamic behavior of the **OPNO** ligand in the TYR active site, with probably several species in equilibrium such as the κ^1-**OPNO** and μ-1,4-**OPNO** adducts.

5. *HOPNO derivatives*

Naturally occurring aurone compounds have been reported in the literature as inhibitors in melanin biosynthesis in human melanocytes.[118] The true effect of aurones is controversial since some compounds of this family act as alternative substrates rather than inhibitors.[119,120] Nevertheless, aurones display strong affinities for TYR, which encourages their integrations in TSA moieties such as **HOPNO**. Therefore, several **HOPNO**-embedded compounds have been reported in the literature for TYR inhibition.[104,119,121]

The **HOPNO**-embedded aurones **I–III** were evaluated for the inhibition of three forms of TYR, one mammalian (*Hs*TYR), one bacterial (*Sa*TYR) and one fungal (*Ab*TYR) (Table 3). **HOPNO**-embedded aurones **I–III** exhibit better inhibition properties than the parent compound **HOPNO**, although the effect is less pronounced for *Ab*TYR. In the case of *Sa*TYR, the substitution of the aurone moieties seems does not significantly modify the inhibition properties since the three compounds **I–III** exhibit K_I in the range 0.1 μM. On the contrary for the *Hs*TYR, this effect is more important, with a $K_I = 0.35$ μM for compound **I** and $K_I = 1.20$ μM. for compound **III**.

The **HOPNO**-embedded aurones **I–III** were also tested in the inhibition of melanin biosynthesis in a human-integrated cellular model (MNT-1

Table 3. Inhibition constant K_I (μM) on purified TYRs of **HOPNO**-embedded aurones **I–III**.

Inhibitors	HsTYR			SaTYR			AbTYR		
	K_I	type	Ref.	K_I	type	Ref.	K_I	type	Ref.
HOPNO	128	C	[104]	7.7	C	[103]	1.8	C	[97]
I	0.35	C	[104]	0.13	C	[122]	$K_{IC} = 1.27$ $K_{IU} = 1.62$	M	[119]
II	1.02	C	[104]	0.15	C	[122]	$K_{IC} = 0.34$ $K_{IU} = 0.90$	M	[122]
III	1.20	C	[104]	0.16	C	[122]	$K_{IC} = 2.9$ $K_{IU} = 2.5$	M	[122]

Table 4. IC$_{50}$ (μM) of **HOPNO**-embedded aurones **I–III** in TYR inhibition in human MNT1 melanoma cells.[104]

Inhibitors	Purified HsTYR	Lysate	Whole cells	Cytotoxicity
HOPNO	128	1300	150	> 200
I	0.35	16.6	85.3	> 500
II	1.02	30	120	80
III	1.2	34	119	> 500

cells). Results from MNT-1 cell lysates confirmed the tendency observed with purified HsTYR (**I** > **II** ~ **III** >> **HOPNO**). However, they displayed a low capacity to reduce melanogenesis in MNT-1 whole cells. Except compound **II**, the others exhibit a low cytotoxicity (Table 4).

HOPNO-embedded aurones have a great inhibition potency. However, the lower efficiency in suppressing melanogenesis in MNT-1 human whole cells remains to be addressed.

VI. Conclusions

From the medicinal point of view, the TYR TSA compounds are promising inhibitors because they target the binuclear copper active site, which is unique in humans. Their potential lies in the fields of antifungals, anti-bacterials, and in diseases linked to disorders of human melanocytes. Among the TSA compounds, **HOPNO** derivatives present the highest potential due to the possibility of decorating the pyrimidone ring to improve inhibitory properties, cell penetration, and inter-species selectivity.

From the mechanistic point of view, TSA compounds are molecular tools capable of providing functional information on TYR. On one hand, functional studies on the mechanism of TYRs converge on a consensus in which the phenolate binds to the Cu_A site of the $\mu\text{-}\eta^2\text{:}\eta^2$-peroxo species to trigger the O-atom transfer reaction (Figure 7A and B). On the other hand, all the 3D structures of the TSA inhibitors (L-mimosine, kojic acid, and tropolone) with TRP1 and AbTYR show that the inhibitors do not bind to the binuclear copper active site but are positioned through π-stacking and H-bonds interactions. This raises the following question: is it possible that the O-atom transfer takes place without the phenolate being bound to the Cu_A (Figure 23)? This is an interesting question that would deserve to be addressed.

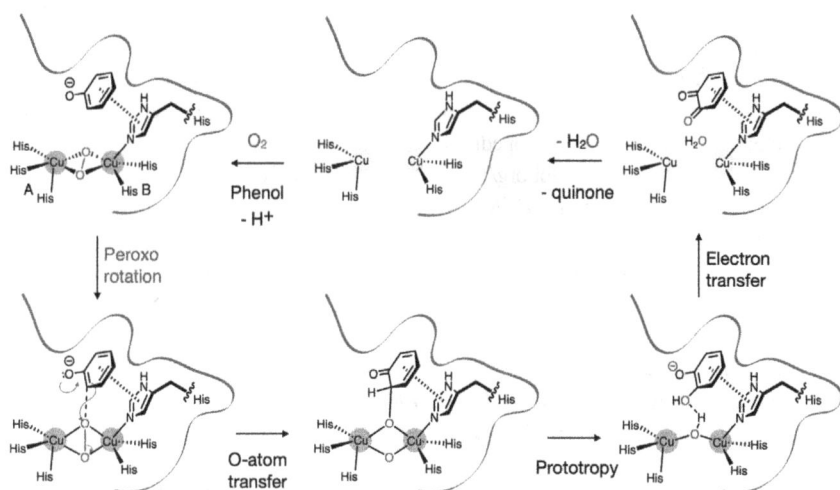

Figure 23. O-atom transfer to phenolate without binding to copper ions.

Acknowledgment

The authors gratefully acknowledge the Cosmethics project, an "Investissements d'Avenir" program (ANR-15-IDEX-02) and the French national synchrotron facility Soleil (proposal number 20120202).

VII. References

1. Cordero, Casadevall, R. J. B.; Melanin, A. *Curr. Biol.* **2020**, R135–R158.
2. Solano, Melanins, F. *New J Sci.* **2014**, 1–28.
3. d'Ischia, M. *et al.* *Pigment Cell Melanoma Res.* **2013**, *26*, 616–633.
4. d'Ischia, M. *et al.* *Pigment Cell Melanoma Res.* **2015**, *28*, 520–544.
5. Hong, L.; Simon, J. D. *J. Phys. Chem.* **2007**, *111*, 7938–7947.
6. Chatelain, M.; Gasparini, J.; Jacquin, L.; Frantz, A. *Biol. Lett.* **2014**, *10*, 20140164.
7. McGraw, K. J. *Oikos.* **2003**, *102*, 402–406.
8. Butler, M. J.; Day, A. W. *Can. J. Microbiol.* **1998**, *44*, 1115–1136.
9. Kaintz, C.; Mauracher, S. G.; Rompel, A. *Adv. Protein Chem. Str.* **2014**, *97*, 1–35.
10. Belle C. Catechol Oxidase and Tyrosinase. In: Encyclopedia of Metalloproteins, Kretsinger, Robert H.; Uversky, Vladimir N.; Permyakov, Eugene A. (Eds.). Springer-Verlag Berlin Heidelberg **2013**, 574–579.
11. Bijelic A., Rompel A. and Belle C. Tyrosinases: enzymes, models and related applications, In Series on Chemistry, Energy and the environment: Bioinspired chemistry, from enzymes to synthetic models, Volume 5, M. Réglier (Ed), World Scientific. **2019**, *5*, 155–183.
12. Tepper, A.W.J.W., Lonardi, E., Bubacco, L. and Canters, G.W. Structure, Spectroscopy, and Function of Tyrosinase; Comparison with Hemocyanin and Catechol Oxidase. In Encyclopedia of Inorganic and Bioinorganic Chemistry, R.A. Scott (Ed.) **2011** (https://doi.org/10.1002/9781119951438.eibc0683)
13. Solomon, E. I.; Sundaram, U. M.; Machonkin, T. E. *Chem. Rev.* **1996**, *96*, 2563–2606.
14. Klabunde, T.; Eicken, C.; Sacchettini, J. C.; Krebs, B. *Nat. Struct. Biol.* **1998**, *5*, 1084–1090.
15. Prexler, S. M.; Frassek, M.; Moerschbacher, B. M.; Dirks-Hofmeister, M. E. *Angew. Chem. Int. Ed.* **2019**, *58*, 8757–8761.
16. Volbeda, A.; Hol, W. G. J. *J. Mol. Biol.* **1989**, *209*, 249–279.
17. Matoba, Y.; Kumagai, T.; Yamamoto, A.; Yoshitsu, H.; Sugiyama, M. *J. Biol. Chem.* **2006**, *281*, 8981–8990.
18. Kaintz, C.; Mauracher, S. G.; Rompel, A. *Adv. Protein Chem. Str.* **2014**, *97*, 1–35.
19. Fujieda, N. *et al.* *J. Biol. Chem.* **2013**, *288*, 22128–22140.
20. Ismaya, W. T. *et al.* *Biochemistry.* **2011**, *50*, 5477–5486.

21. Mauracher, S. G.; Molitor, C.; Al-Oweini, R.; Kortz, U.; Rompel, A. *Acta Cryst.* **2014**, *70*, 2301–2315 ().
22. Pretzler, M.; Bijelic, A.; Rompel, A. *Acta Cryst.* **2017**, 7:1810, 1–10.
23. Bijelic, A.; Pretzler, M.; Molitor, C.; Zekiri, F.; Rompel, A. *Angew. Chem. Int. Ed. Engl.* **2015**, *54*, 14677–14680.
24. Klabunde, T.; Eicken, C.; Sacchettini, J. C.; Krebs, B. *Nat. Struct. Biol.* **1998**, *5*, 1084–1090.
25. Sendovski, M.; Kanteev, M.; Ben-Yosef, V. S.; Adir, N.; Fishman, A. *J. Mol. Biol.* **2011**, *405*, 227–237.
26. Li, Y.; Wang, Y.; Jiang, H.; Deng, J. *Proc. Nat. Acad. Sci.* **2009**, *106*, 17002–17006.
27. Masuda, T.; Momoji, K.; Hirata, T.; Mikami, B. *FEBS J.* **2014**, *281*, 2659–2673.
28. Lai, X.; Soler-López, M.; Wichers, H. J.; Dijkstra, B. W. *Plos One.* **2016**, *11*, e0161697.
29. Favre, E.; Daina, A.; Carrupt, P.-A.; Nurisso, A. *Chem. Biol. Drug Design.* **2014**, *84*, 206–215.
30. Lai, X.; Wichers, H. J.; Soler-López, M.; Dijkstra, B. W. *Chem. Eur. J.* **2017**, *24*, 47–55.
31. Jumper, J. *et al. Nature* **2021**, *596*, 583–589.
32. Zou, C. *et al. Molecules.* **2017**, *22*, 1836–1847.
33. Matoba, Y.; Yoshitsu, H.; Jeon, H.-J.; Oda, K.; Noda, M.; Kumagai, T.; Sugiyama, M. **2008**, PDB. doi: 10.2210/pdb2ZMY/pdb.
34. Bubacco, L.; Spinazze, R.; Longa, S.; Della, Benfatto, M. *Arch. Biochem. Biophys.* **2007**, *465*, 320–327.
35. Mirica, L. M. *et al. J. Am. Chem. Soc.* **2006**, *128*, 2654–2665.
36. Mirica, L. M.; Ottenwaelder, X.; Stack, T. D. P. *Chem. Rev.* **2004**, *104*, 1013–1045.
37. Nasir, M. S.; Cohen, B. I.; Karlin, K. D. *J. Am. Chem. Soc.* **1992**, *114*, 2482–2494.
38. Yamazaki, S.-I.; Itoh, S. *J. Am. Chem. Soc.* **2003**, *125*, 13034–13035.
39. Rolff, M.; Decker, H.; Tuczek, F. *Chem. Soc. Rev.* **2011**, *40*, 4077–22.
40. Decker, H.; Schweikardt, T.; Tuczek, F. *Angew. Chem. Int. Ed. Eng.* **2006**, *45*, 4546–4550.
41. Ismaya, W. T. *et al. Acta Cryst.* **2011**, *67*, 575–578.
42. Fujieda, N. *et al. Angew. Chem. Int. Ed. Eng.* **2020**, *59*, 13385–13390.
43. Gautron, A. *et al. Pigm. Cell Melanoma R.* **2021**, *34*, 836–852.
44. Lai, X.; Wichers, H. J.; Soler-López, M.; Dijkstra, B. W. *Angew. Chem. Int. Ed. Eng.* **2017**, *56*, 9812–9815.
45. Dolinska, M. B.; Wingfield, P. T.; Young, K. L.; Sergeev, Y. V. *Pigment Cell Melanoma Res.* **2019**, *32*, 753–765.
46. Ishikawa, M.; Kawase, I.; Ishii, F. *Biol. Pharm. Bull.* **2007**, *30*, 677–681.
47. Bastonini, E.; Kovacs, D.; Picardo, M. *Ann. Dermatol.* **2016**, *28*, 279–289.
48. Boyle, A. L.; Boyle, J. L.; Haupt, H. M.; Stern, J. B.; Multhaupt, H. A. B. *Arch. Pathol. Lab. Med.* **2002**, *126*, 816–822.

49. Weinstein, D.; Leininger, J.; Hamby, C.; Safai, B. *J. Clin. Aesthet. Dermatol.* **2014**, *7*, 13–24.
50. World Cancer Research Fund International <https://www.wcrf.org/cancer-trends/skin-cancer-statistics/>
51. Brożyna, A. A.; Jóźwicki, W.; Carlson, J. A.; Slominski, A. T. *Hum. Pathol.* **2013**, *44*, 2071–2074.
52. Rezaei, T. *et al. Pigment Cell Melanoma Res.* **2021**, *34*, 869–891.
53. Jawaid, S.; Khan, T. H.; Osborn, H. M. I.; Williams, N. A. O. *Anticancer Agents Med. Chem.* **2009**, *9*, 717–727.
54. Vargas, A. J. *et al. Integr. Cancer Ther.* **2011**, *10*, 328–340.
55. Nishimura, M. I.; Al-Khami, A. A.; Mehrotra, S.; Wolfel, T. *Cancer Therapeutic Targets*, Ed. J. L.; Marshall J. L.; Springer, New York, **2016**, 1–8.
56. Cabaço, L. C.; Tomás, A.; Pojo, M.; Barral, D. C. *Front. Oncol.* **2022**, *12*, 887366.
57. Buitrago, E. *et al. Curr. Top. Med. Chem.* **2016**, *16*, 3033–3047.
58. Roulier, B.; Pérès, B.; Haudecoeur, R. *J. Med. Chem.* **2020**, *63*, 13428–13443.
59. Slominski, R. M. *et al. Front. Oncol.* **2022**, *12*, 842496.
60. Rendon, M.; Horwitz, S. *Ann. Dermatol. Vénéréol.* **2012**, *139*, S153–S158.
61. Desai, S. *Treatment of Skin Disease Comprehensive. Therapeutic Strategies (5th edition)*, Eds. Lebwohl, M. G.; Heymann, W. R.; Berth-Jones, J.; Elsevier, **2017**, 658–661.
62. Ogden, S.; Griffiths, C. E. M. *Treatment of Skin Disease Comprehensive. Therapeutic Strategies (5th edition)*, Eds. Lebwohl, M. G.; Heymann, W. R.; Berth-Jones, J.; **2017**, Elsevier, 493–495.
63. Qian, W. *et al. Exp. Ther. Med.* **2020**, *20*, 173–185.
64. Pillaiyar, T.; Manickam, M.; Namasivayam, V. *J. Enzym. Inhib. Med. Chem.* **2020**, *32*, 403–425.
65. Burki, T. *Lancet Diabetes Endocrinol.* **2020**, *9*, 10.
66. Desmedt, B. *et al. J. Eur. Acad. Dermatol.* **2016**, *30*, 943–950.
67. Loizzo, M. R.; Tundis, R.; Menichini, F. *Compr. Rev. Food Sci. Food Saf.* **2012**, *11*, 378–398.
68. Gębalski, J.; Graczyk, F.; Załuski, D. *J. Enzym. Inhib. Med. Chem.* **2022**, *37*, 1120–1195.
69. Vaezi, M. *J. Biomol. Struct. Dyn.* **2022**, 1–13.
70. Peng, Z. *et al. Crit. Rev. Food Sci.* **2022**, *62*, 1–42.
71. Obaid, R. J. *et al. RSC Adv.* **2021**, *11*, 22159–22198.
72. Hariri, R.; Saeedi, M.; Akbarzadeh, T. *J. Pept. Sci.* **2021**, *27*, e3329.
73. Riaz, R. *et al. Mini Rev. Org. Chem.* **2021**, *18*, 808–828.
74. Zhang, X. *et al. J. Enzym. Inhib. Med. Chem.* **2021**, *36*, 2104–2117.
75. Bonesi, M. *et al. Curr. Med. Chem.* **2019**, *26*, 3279–3299.
76. Mendes, E.; Perry, M.; de J.; Francisco, A. P. *Expert Opin. Drug Discov.* **2014**, *9*, 533–554.

77. Kampatsikas, I.; Bijelic, A.; Pretzler, M.; Rompel, A. *Sci. Rep.* **2017**, *7*, 8860.
78. Criton, M.; Mellay-Hamon, V. L. *Bioorg. Med. Chem. Letters* **2008**, *18*, 3607–3610.
79. Yi, W. *et al. Chem. Pharm. Bull.* **2010**, *58*, 752–754.
80. Yi, W. *et al. Chem. Pharm. Bull.* **2009**, *57*, 1273–1277.
81. Yi, W. *et al. Eur. J. Med. Chem.* **2011**, *46*, 4330–4335.
82. Hałdys, K.; Latajka, R. *Med. Chem. Comm.* **2019**, *10*, 378–389.
83. Liu, J.; Yi, W.; Wan, Y.; Ma, L.; Song, H. *Bioorg. Med. Chem.* **2008**, *16*, 1096–1102.
84. Liu, J. *et al. Eur. J. Med. Chem.* **2009**. *44*, 1773–1778.
85. Li, Z.-C. *et al. J. Agri. Food Chem.* **2010**, 2–5.
86. Xue, C.-B. *et al. Bioorg. Med. Chem.* **2007**, *15*, 2006–2015.
87. Thanigaimalai, P.; Hoang, T. A. L.; Lee, K.-C.; Kim, Y.; Jung, S.-H. *Bioorg. Med. Chem. Lett.* **2010**, *20*, 2991–2993.
88. Pan, Z.-Z. *et al. J. Agri. Food Chem.* **2012**, *60*, 10784–10788.
89. Buitrago, E. *et al. Inorg. Chem.* **2014**, *53*, 12848–12858.
90. Tudela, J.; Lozano, J. A.; Garcia-Canovas, F. *Phytochemistry.* **1987**, *26*, 917–919.
91. Kyriakou, S. *et al. Invest. New Drug.* **2020**, *38*, 621–633.
92. Chen, J. S. *et al. J. Agr. Food Chem.* **1991**, 39, 1396–1401.
93. Chen, J. S.; Wei, C. I.; Marshall, M. R. *J. Agr. Food Chem.* **1991**, *39*, 1897–1901.
94. Espín, J. C.; Wichers, H. J. *J. Agr. Food. Chem.* **1999**, *47*, 2638–2644.
95. Takahashi, S. *et al. Bioorg. Med. Chem.* **2010**, *18*, 8112–8118.
96. Yoshimori, A. *et al. Bioorg. Med. Chem.* **2014**, *22*, 6193–6200.
97. Peyroux, E. *et al. Inorg. Chem.* **2009**, *48*, 10874–10876.
98. Orio, M. *et al. Chem. Eur. J.* **2011**, *17*, 13482–13494.
99. Goliênik, M.; Stojan, J. *Biochem. Mol. Biol. Edu.* **2004**, *32*, 228–235.
100. Fogal, S. *et al. Mol. Biotechnol.* **2014**, *57*, 45–57.
101. Nesterov, A. *et al. Chem. Pharm. Bull.* **2008**, *56*, 1292–1296.
102. Deri, B. *et al. Sci. Rep.* **2016**, *6*, 34993–10.
103. Bochot, C. *et al. Chem. Eur. J.* **2013**, *19*, 3655–3664.
104. Haudecoeur, R. *et al. ACS Med. Chem. Lett.* **2017**, *8*, 55–60.
105. Bubacco, L.; Gastel, M.; van, Groenen, E. J. J.; Vijgenboom, E.; Canters, G. W. *J. Biol. Chem.* **2003**, *278*, 7381–7389.
106. Gastel, M.; van, Bubacco, L.; Groenen, E.; Vijgenboom, E.; Canters, G. W. *FEBS Lett.* **2000**, *474*, 228–232.
107. Benfatto, M.; Della Longa, S.; Pace, E.; Chillemi, G.; Padrin, C.; Natoli, C. R.; Sanna, N. *Comput. Phys. Commun.* **2021**, *265*, 107992.
108. Bochot, C. *et al. Chem. Comm.* **2014**, *50*, 308–310.
109. Torelli, S. *et al. Inorg. Chem.* **2000**, *39*, 3526–3536.
110. Buitrago, E. *et al. Chem. Eur. J.* **2021**, *27*, 4384–4393.
111. Lee, Y. S.; Park, J. H.; Kim, M. H.; Seo, S. H.; Kim, H. J. *Arch. Pharm.* **2006**, *339*, 111–114.
112. Kahn, V.; Andrawis, A. *Phytochemistry.* **1985**, *24*, 905–908.

113. Espin, J. C.; Wichers, H. J. *J. Agri. Food Chem.* **1999**, *47*, 2638–2644.
114. Yoshimori, A. *et al. Bioorg. Med. Chem.* **2014**, *22*, 6193–6200.
115. Takahashi, S. *et al. Bioorg. Med. Chem.* **2010**, *18*, 8112–8118.
116. "Bioinspired Chemistry: From Enzymes to Synthetic Models.", Ed. M. Réglier, In "Series in Chemistry, Energy and Environment", Eds. K. Kadish and R. Guilard, World Scientific. Vol. 5: 2019.
117. Addison, A. W.; Rao, T. N.; Reedijk, J.; van Rijn, J.; Verschoor, G. C. *J. Chem. Soc. Dalton Trans.* **1984**, 1349–1356.
118. Okombi, S. *et al. J. Med. Chem.* **2006**, *49*, 329–333.
119. Dubois, C. *et al. Chem. Bio. Chem.* **2012**, *13*, 559–565.
120. Marková, E. *et al. J. Agri. Food Chem.* **2016**, *64*, 2925–2931.
121. Haudecoeur, R. *et al. Chem. Bio. Chem.* **2014**, *15*, 1325–1333.
122. Dubois, C. La tyrosinase: étude de nouveaux effecteurs, Aix-Marseille Université, Oct. **2012**.

3 Modeling Tyrosinase Activity Using *m*-Xylyl-Based Ligands: Ring Hydroxylation, Reactivity, and Theoretical Investigation

Puneet Gupta[¶] and Rabindranath Mukherjee[†,‡,§]

[¶]Department of Chemistry, Indian Institute of Technology Roorkee, Roorkee, Uttarakhand 247 667, India
[†]Department of Chemistry, Indian Institute of Technology Kanpur, Kanpur, Uttar Pradesh 208 016, India
[‡]Present address: Department of Chemistry and Chemical Biology, Indian Institute of Technology (Indian School of Mines) Dhanbad, Dhanbad, Jharkhand 826 004, India
[§]rnm@iitk.ac.in

I. Introduction

A. General Considerations

Dioxygen is crucial for aerobic organisms, serving as a primary source of energy with its thermodynamically favorable reduction to water. It is used as an oxidant in nature. However, the use of dioxygen as an oxidant is not straightforward. Dioxygen is a diradical species and hence its triplet ground state electronic structure makes it unreactive, under ambient conditions, toward most organic substrates, which are generally diamagnetic and hence have singlet ground state. This is due to spin-forbidden nature of the reaction.[1,2] In nature, metalloenzymes activate O_2 and its further processing is done by reduction for oxidation reactions.[3] The reduced forms of O_2 such as O_2^{2-} (peroxide) or O^{2-} (oxide) exist in singlet ground states, therefore they can easily react with singlet organic molecules. In addition, O atom incorporation into biological substrates occurs by functionalization through mono- or dioxygenation processes mediated by metalloenzymes. The corresponding metalloenzymes are called monooxygenases and dioxygenases, respectively. The former one incorporates one O atom to the substrate and the other O atom of dioxygen is reduced to H_2O. The latter one incorporates both the O atoms to the substrate.

Molecular-level understanding of basic chemical principles and mechanisms of O_2 activation at metal sites and its subsequent processing for oxygenations and oxidations of organic molecules have been of interest for a few decades now. The continued interests are because of their potential relevance to biological processes.[4–16] Due to its generally accessible Cu(II) and Cu(I) states and bioavailability, copper plays a wide variety of roles in nature.[1a,4]

In the domain of synthetic modeling approach involving biomimetic studies,[17] many factors affect the reactivity of low-molecular-weight metal complexes with tailor-made organic ligands toward dioxygen.[1–16,17b] Since the reaction causes one-electron oxidation of the metal center with reduction of dioxygen, the reduction potential of M^{II}/M^{I} (for M = Cu) or M^{III}/M^{II} (for M = Fe) redox couples is one of the important factors governing the reaction between the metal complex and dioxygen.[9c] The redox potentials of metal complexes vary dramatically depending upon the ligand environment. The solvent exerts a strong effect as well. One-electron reduction of dioxygen is a thermodynamically unfavorable process, since for $O_2 + 1e^- = O_2^-$ redox process the $E^o = -0.35$ V *versus* NHE at pH 7. However, for the two-electron, $O_2 + 2H^+ + 2e^- = H_2O_2$ reduction process ($E^o = 0.28$ V *vs.* NHE at pH 7) and for four-electron, $O_2 + 4H^+ + 4e^- = 2H_2O$ reduction process ($E^o = 0.82$ V *vs.* NHE at pH 7) are favorable reactions. Although the reduction potential is a good predictor of the reactivity of the metal complexes toward dioxygen, it is not directly correlated with the feasibility for the formation of the dioxygen adduct, since the redox potentials are defined for outer-sphere-type electron-transfer reactions.[9c] On the other hand, metal-dioxygen interaction is an inner-sphere-type electron transfer process.

The feasibility of formation of the "$M-O_2$" or "$M-O_2-M$" adduct is associated with both thermodynamic and kinetic factors.[9c] The former factor is directly related to the stability of the resultant metal-dioxygen adduct. The contributors to stability of the metal complex due to the ligand environment include the ligand donor set, geometry, and substituent-induced electronic and steric effects, and chelate-ring size.[17b] The availability of an open coordination site at the metal center to bind dioxygen and the steric bulk of the ligand substituent groups are also important, having a striking kinetic effect on the binding rate of dioxygen to the metal center.

B. Scope of the Review

Copper proteins/enzymes participate in activities such as dioxygen binding, dioxygen activation, and oxidation and reduction processes of the substrates.[4] The activities of copper enzymes triggered by dioxygen can be broadly classified as oxidase, monooxygenase, or dioxygenase.[5a,15] These reactivity aspects have long interested inorganic chemists. The protein of interest in the context of this article is that of antiferromagnetically coupled dicopper-containing monooxygenase, tyrosinase, in its oxygenated state.[1-14,17a] The structural and spectroscopic modeling of tyrosinase (monophenolase and diphenolase activity; see below) active site is an important and challenging topic of bioinorganic chemistry of copper. It has been well-documented that coordination chemists can make significant contributions to reactivity studies and mechanisms.[1a,2,3,5-17b,c] The closely related members of this family of proteins are hemocyanin and catechol oxidase.[4,9c,d,14] The basic structural and electronic structural properties of these type 3 copper proteins/enzymes[9d,17a] are the necessary reference for any attempt to reproduce in biomimetic systems key aspects of the protein structure and associated reactivity.

A large volume of outstanding contributions has been made in the copper–dioxygen chemistry by synthetic chemists.[1a,2,3,5-16] The success story on the synthetic front in structural and spectroscopic characterization of three Cu_2O_2 cores (side-on peroxo, bis-oxido, and end-on peroxo) have not only enriched but also sharpened our knowledge on dicopper proteins and enzymes. The outcome of an in-depth mechanistic studies of tyrosinase-like reactivity on suitable substrates using these Cu_2O_2 reactive intermediates is considered as a breakthrough in molecular-level understanding of the functioning of tyrosinase.

The focus of this article has been divided into two main parts. The first section describes how ligand modifications[8a,11] can often result in drastically different dioxygen reactivity of the dicopper(I) complexes. We will examine several examples to highlight the ligand aspects such as chelate-ring size, electronic and steric effects, nitrogen donor type, and their consequences on the reactivity of the copper(I) complexes toward O_2. The ligand systems we have primarily paid attention to are m-xylyl-based dinucleating chelating ligands of varying denticity. We will emphasize the importance of such ligands accommodating two copper ions, initially as Cu(I) and after activation of O_2 as Cu(II)/Cu(III), the subsequent reactivity of the Cu_2O_2 intermediates (spectroscopically characterized or even not characterized) toward the aromatic ring to afford the final intramolecular m-xylyl ring hydroxylated product as μ-phenoxo μ-hydroxo-bridged dicopper(II) complex. Examples are also chosen to exhibit general aromatic ring hydroxylation and formation of di-μ-phenoxo-bridged dicopper(II) complex. In the second part of the article we will discuss the electronic structure of the reactive intermediates (Cu_2O_2 adducts) and understanding of their reactivity properties.

It is the purpose of this article to highlight the biomimetic studies on tyrosinase. Our own research efforts have focused on both the aspects. It is admitted that this report is not exhaustive, and interested readers are referred to excellent review articles.[1–16]

II. Dicopper Proteins — Brief Overview

Hemocyanin, tyrosinase, and catechol oxidase belong to the class of type 3 copper proteins.[14c] An overview of the structure and functioning of these dicopper-containing proteins/enzymes is briefly discussed here.

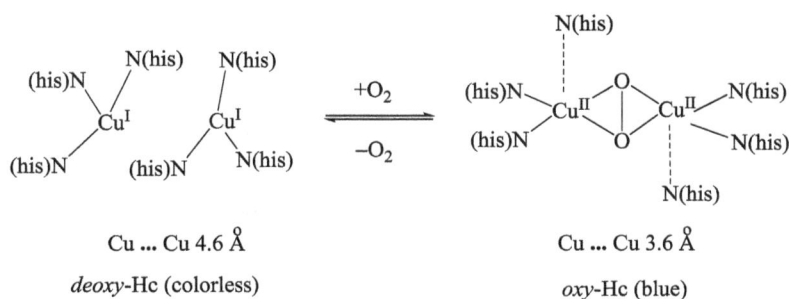

Cu ... Cu 4.6 Å   Cu ... Cu 3.6 Å

deoxy-Hc (colorless)   *oxy*-Hc (blue)

Figure 1.   Schematic drawing of the structures of *deoxy*- and *oxy*-state of Hc.

A. Hemocyanins

Hemocyanins (Hc) are the oxygen-carrier protein of molluscs and arthropods. Hemocyanin, unlike hemoglobin, has no heme group; the copper is bound directly to the protein side chain. Reversible binding of dioxygen is depicted in Figure 1. X-ray structures of both *deoxy*- and *oxy*-forms of Hc have been determined.[4,14b,18]

B. Tyrosinases

Polyphenol oxidase is a generic term for the group of enzymes that catalyze the oxidation of phenolic compounds to produce brown color on cut surfaces of fruits and vegetables.[4,14c,19] The enzyme tyrosinase (Tyr) catalyzes the hydroxylation of monophenols (tyrosine) to *o*-diphenols (L-3,4-dihydroxyphenylalanine, L-DOPA; monophenolase activity) and subsequent two-electron oxidation to *o*-quinones (L-DOPA-quinone; diphenolase activity), which constitutes the first step of melanin biosynthesis through a series of spontaneous, non-enzymatic reactions (Figure 2).

Tyrosinase appears to be ubiquitous in living organisms. It is widely distributed in plants, fungi, bacteria, mammalians, and animals. When potatoes, apples, bananas, sweet potatoes, or mushrooms are injured they turn brown.[4,19] This is due to the conversion of tyrosine to the pigment melanin, by the sequence of reactions shown in Figure 2. The same process causes skin to turn brown, following exposure to ultraviolet radiation. The enzymatic reactions are catalyzed by tyrosinase. The enzyme is present in

Tyrosine   L-DOPA   L-DOPA-quinone

L-DOPA-quinone ⟶ polymeric pigments (melanin)

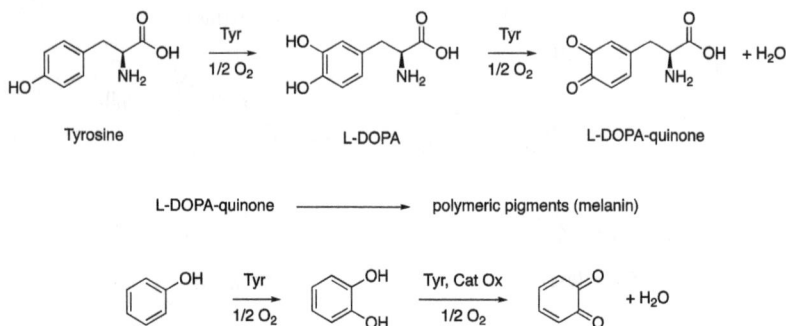

Figure 2. Tyrosinase and catechol oxidase activities.

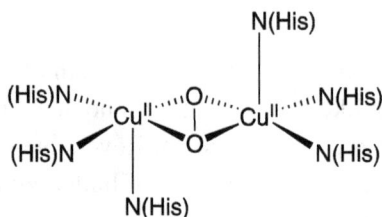

Figure 3. Active site of *oxy*-tyrosinase, depicting a $\mu\text{-}\eta^2\text{:}\eta^2$-peroxo-dicopper(II) core.

the interior of the plant material and, since the reaction requires molecular oxygen, the pigmentation does not occur until the interior is exposed. Since melanin is a key pigment of some phenomena such as suntan, skin disorder, and bruising of fruits, tyrosinase has practical and economic importance, and thus attracting much attention from cosmetology, medicine, and agriculture to control the synthesis of the melanin pigment.[19]

Tyrosinase is an O_2-activating enzyme. The active site of tyrosinase has a dinuclear copper center with each copper coordinated by three histidines. The oxy-state of the enzyme contains a O_2^{2-} ion bound to two copper(II) centers in a η^2-fashion (Figure 3).[19a] The X-ray structure of *oxy*-Tyr is very similar to *oxy*-Hc (Figure 1), the only difference being the active site of Tyr is more exposed to external substrate than in Hc.

C. Catechol Oxidase

Catechol oxidase (Cat Ox) is an enzyme that catalyzes the two-electron oxidation of catechols to *o*-quinones (diphenolase activity/catecholase

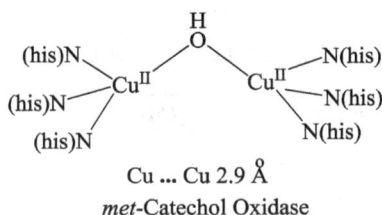

Cu ... Cu 2.9 Å

met-Catechol Oxidase

Figure 4.   Structure of the *met*-form of catechol oxidase.

activity).[14c,20] In contrast to tyrosinase, which performs both the hydroxylation and the oxidation step, catechol oxidase is only able to mediate the
latter reaction (Figure 2), without acting on phenols. Catechol oxidases
are found in plant tissues and in some insects and represent a group of
ubiquitous oxidases in plants. The mechanism of its function is closely
associated with that of tyrosinases. The active site structure of its physi-
ologically inactive *met*-form is shown in Figure 4.

III. Three Cu₂O₂ Core Structures

The X-ray structure of *oxy*-Hc and *met*-form of catechol oxidase are
shown in Figures 1 and 4, respectively. The X-ray structure of *oxy*-Tyr
(Figure 3) is very similar to *oxy*-Hc.

For the dicopper sites, three basic Cu_2O_2 core structures (Figure 5)
arising from the reactions of discrete Cu(I) complexes and O_2 have been
widely investigated by synthetic chemists. One of the highlights of the
success story of synthetic chemists is the structural characterization of
these dicopper cores. A specific Cu_2O_2 core signifies a particular spectroscopic and chemical properties.[1–3,5–16]

Side-on peroxo-dicopper(II)/$\{Cu^{II}_2(\mu\text{-}\eta^2{:}\eta^2\text{-}O_2)\}^{2+}$/$\{\mathbf{Cu_2P^S}\}$ and bis
(μ-oxido)dicopper(III) $\{Cu^{III}_2(\mu\text{-}O)_2\}^{2+}$/$\{\mathbf{Cu_2O_2}\}$ cores show electrophilic
character and can mediate tyrosinase-like phenolate *ortho*-hydroxylation
reactions. End-on *trans*-peroxo-dicopper(II) $\{Cu^{II}_2(\mu\text{-}\eta^1{:}\eta^1\text{-}O_2)\}^{2+}$/
$\{\mathbf{Cu_2P^E}\}$ core exhibits nucleophilic and basic behavior. The tyrosinase-
like reactivity has recently been observed for end-on $\{Cu^{II}_2\text{-}(\mu\text{-}\eta^1{:}\eta^1\text{-}O_2)\}^{2+}$
core as well (see below). It should be noted that in $\{\mathbf{Cu_2P^S}\}$ and $\{\mathbf{Cu_2P^E}\}$
binding modes the copper centers are in +2 oxidation state and O_2 is in
its two-electron reduced form (O_2^{2-}); whereas in $\{\mathbf{Cu_2O_2}\}$ mode both

Distances (Å)	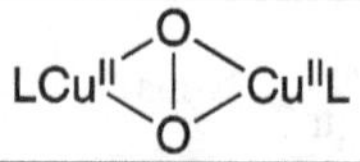 $LCu^{II}\diagup^{O}_{O}\diagdown Cu^{III}L$	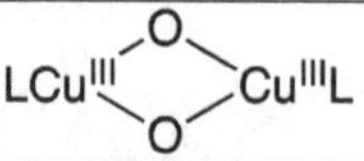 $LCu^{III}\diagup^{O}_{O}\diagdown Cu^{III}L$	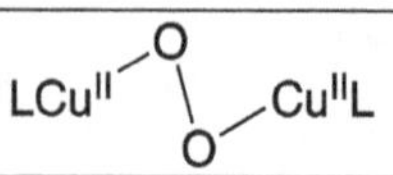 $LCu^{II}\diagup^{O}_{O}\diagdown Cu^{III}L$
	{Cu_2P^S}	{Cu_2O_2}	{Cu_2P^E}
$d(Cu{\cdots}Cu)$	3.51	2.80	4.36
$d(O–O)$	1.42	2.32	1.43
$d(Cu–O)$	1.92	1.82	1.85

Figure 5. Three isoelectronic Cu_2O_2 binding modes, their shorthand notations, and geometrical parameters. L: supporting ligands.

coppers are in +3 state and O_2 is in its four-electron reduced (O^{2-}) form. The three motifs are electronic isomers.

Excellent review articles are available in the literature on molecular, structural, spectroscopic, electronic structural, and reactivity aspects of these three cores.[1–3,5–16] In-depth discussion on the mechanism of the functioning of tyrosinases[1–16,19] and catechol oxidases[20] are available in the literature. The intricate structure-directed functional properties of three dicopper-containing proteins/enzymes Hc, Tyr, and Cat Ox are noteworthy.

IV. Biomimetic Studies on Tyrosinase

A. Intramolecular *m*-Xylyl and Aromatic Ring Hydroxylation

Fascinated by the biological role of tyrosinase, several synthetic models have so far been made.[2,5–11,15,16,21–26] The reactions of Cu(I) complexes of simple monodentate/bidentate and tailor-made chelating ligands with O_2 and the oxidative properties of the resulting Cu_n–O_2 species has attracted much interest over the past several decades. Excellent review articles have appeared in the literature, which has addressed various aspects of this field.[1–16,19,21–24] Due to the substitution lability of Cu(I) and Cu(II) ions, the ligand coordination subtly controls which species is formed and their stability.[8a] In fact, the ligand architecture defines the Cu_n–O_2 structure to be attained during oxygenation and subsequent reactivity pattern. It has been well documented that even seemingly minor alterations in the ligand dramatically affect the oxygenation reactions, thus providing a direct mechanistic probe.[8a]

In what follows, we highlight some of the biomimetic studies aimed at generating and characterizing (structurally and/or spectroscopically) the copper–dioxygen intermediates and subsequent reactions of such reactive species to bring about hydroxylation/oxidation of copper-bound ligand and/or of externally added substrates of relevance to tyrosinase activity. Also included are the systems where ligand hydroxylation/oxidation has occurred, as revealed by X-ray crystallographic and/or spectroscopic analysis of the final product, even though direct proof of the formation of copper–dioxygen intermediate could not be provided. Apart from *m*-xylyl-based ligands a few systems that had undergone aromatic ring hydroxylation/oxidation by reactive copper–dioxygen intermediate have also been considered.

B. Intramolecular *m*-Xylyl and Aromatic Ring Hydroxylation

The earliest examples of biomimetic oxygenation reactions (tyrosinase-like activity) came from the outstanding studies performed by Karlin's group on aromatic hydroxylation reaction occurring in the *m*-xylyl-based ligand (Figure 6), upon oxygenation of dicopper(I) complexes.[5,11,21] Moreover, the most significant insights into the chemical activation of dioxygen by dinuclear copper(I) sites, where the two steps of dioxygen binding and ligand oxygenation were separately identified.[21b]

Specifically, Karlin's group reported the synthesis of a dinucleating ligand system (**1**: Scheme 1), which provides three nitrogen donors ((2-pyridyl)ethylamine unit to each copper ion), separated by a *m*-xylyl spacer.[5,21] Upon reaction with O_2 at low temperature (193 K), the dioxygen adducts form reversibly and these subsequently yield 2-xylylene-hydroxylated products, which are phenoxo- and hydroxo-bridged dicopper(II) complexes (Scheme 1). The products have been characterized *via* the X-ray structure of the complex with R = H and by their UV–visible

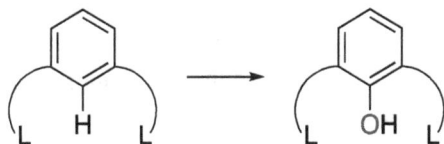

Figure 6.   Schematic representation of *m*-xylyl ring hydroxylation.

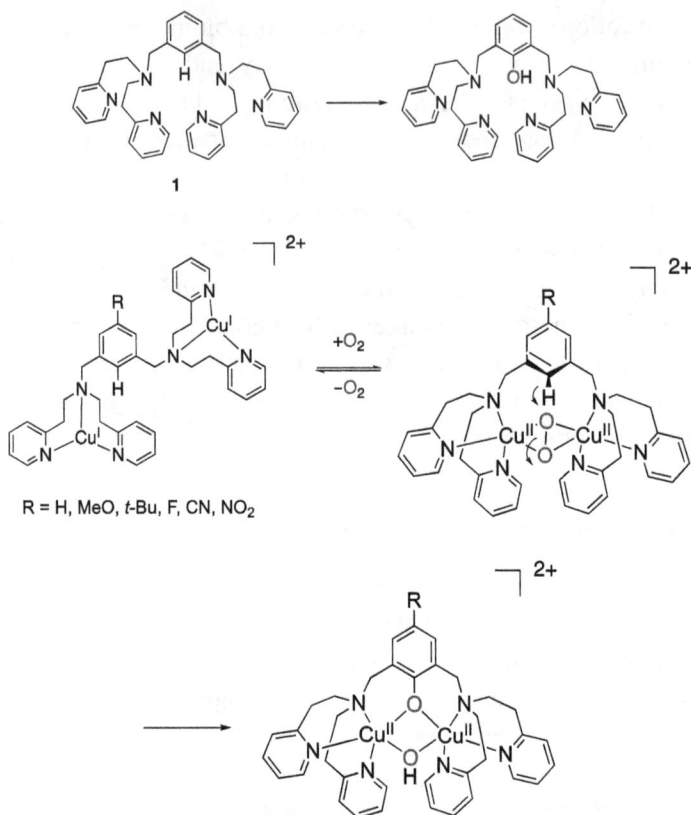

Scheme 1. A copper monooxygenase (aromatic *m*-xylyl ring hydroxylation) model system of Karlin and coworkers.

spectral features. Like monooxygenases, one O atom of O_2 is inserted into C–H bond to form C–OH (Figure 6) and the other O atom is reduced to OH^- ion. The involvement of the side-on dicopper(II)-peroxo intermediate $\{Cu^{II}_2(\mu\text{-}O_2)\}^{2+}$ $\{\mathbf{Cu_2P^S}\}$, formed upon reaction of dicopper(I) complex of the *m*-xylyl ligand and dioxygen, was firmly established.[21b,c]

Mechanistically, electrophilic attack (see below) of the intermediate was proved by the substituent effect on the 2- and 5-positions of the *m*-xylyl ring.[21b,d] For this ligand system, coordination of the Cu^{II}/Cu^{I} ions by three N donors at each arm gives rise to the formation of six-membered chelate rings. This confers stability not only to Cu(I) ions but also allows the Cu(II) ions to attain tetragonal geometry. Karlin's Cu_2–O_2-mediated

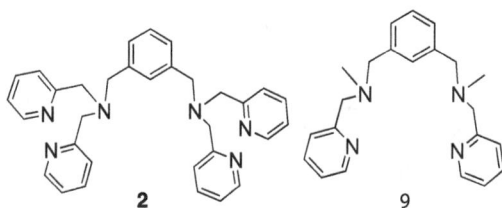

Figure 7. The ligands that failed to demonstrate intramolecular *m*-xylyl ring hydroxylation.

intramolecular *m*-xylyl ring hydroxylation reaction (Scheme 1) occurs due to electrophilic attack (for details see Section 1.5.3.1) of {$\mathbf{Cu_2P^S}$} species.

A small perturbation of ethylpyridine to methylpyridine arm in the ligand system (Figure 7), with six-membered (**1**) to five-membered (**2**) chelate ring formation, leads to irreversible oxidation of Cu(I) to bis(μ-hydroxo)dicopper(II) complex, that is, without xylyl-hydroxylation, following a drastic change in the path of Cu(I)–O_2 reactivity (4Cu(I) + O_2 + 4H$^+$ = 4Cu(II) + 2H$_2$O).[5e]

Inspired by the success story of Karlin's aromatic ring hydroxylation reaction,[5,11,21] systematic modifications of the original *m*-xylyl ligand system (**1**) have been done by many researchers.[22–25] The ligand variations include:

(i) tridentate arms with macrocyclic N_3 donor
(ii) tridentate N_3 five-membered chelate ring (methyl substituents near donor site of the pyridyl rings) (steric effect)
(iii) N_3 donor with one (2-pyridyl)ethylamine and one (2-pyridyl)methylamine arm
(iv) unsymmetrical arms with one tridentate and the other bidentate
(v) tridentate N_3 to bidentate N_2 arm (two to one (2-pyridyl)ethylamine arm) of non-Schiff base and Schiff base variety
(vi) bidentate Schiff base variety with other donors instead of pyridine
(vii) replacing heterocycles by aliphatic amines of non-Schiff base and Schiff base variety
(viii) macrocyclic systems (two N donor as well as three N donor arms) with Schiff base and non-Schiff base variety.

The protocol followed by other groups for intramolecular aromatic *m*-xylyl ring hydroxylation reactions is like that of Karlin and coworkers.[11,21] Dissolving the isolated yellowish dicopper(I) complex, obtained from the reaction between the ligand and appropriate copper(I) salt in suitable solvents (CH_2Cl_2, MeOH, MeCN, acetone, DMF, or THF), under anaerobic conditions, or anaerobic solution–generated dicopper(I) complex in solution, followed by exposure to dry O_2 at low temperatures (183–213 K) or even at room temperature (298 K). Attempts were made to spectroscopically identify the intermediate formed, with characteristic absorption and resonance Raman spectroscopic signatures,[5,11,21] at low temperatures and then warming the solutions to room temperature. Efforts have also been made to spectroscopically characterize the final green solutions and/or isolate the solid in pure form for its X-ray crystallographic structural analysis. Green solutions of *μ*-phenoxo *μ*-hydroxo-bridged dicopper(II) complexes display phenoxo-to-Cu(II) and hydroxo-to-Cu(II) ligand-to-metal charge-transfer (LMCT) transition at ~350 nm.

Ligand variation types (i) and (ii) (see above) afforded **3** and **4**, respectively,[25a,b] and (iii) and (iv) resulted in the isolation of **5** and **6**, respectively,[25c,d] and (v) afforded **7** and **8**, respectively[25c,e,f] (Figure 8). Changing a six-membered chelate ring forming ligand **8** to a five-membered chelate ring forming ligand **9** (Figure 7) notably stops intramolecular ring hydroxylation.[25f] Similar behaviors were observed for ligands **2** and **9**. Ligand variation (v) afforded **10**.[25g] Ligand variation type (vi) led to the generation of species **11–17**.[25h–l] The presence of six-membered chelate rings seems crucial for the stability of binuclear copper(I) counterparts of **12–14**, without phenolic group, since all attempts to prepare dicopper(I) complexes containing five-membered chelate rings were unsuccessful.[25i,j] Generation of **13** was also accomplished independently,[25l] which belongs to this category. Ligand variation type (vii) led to the isolation of **18–20**[25m–o] and **21**.[25p]

The ligand design variation type (viii) was considered to change from open-chain to macrocylic *m*-xylyl-based systems. The 18-atom tetra Schiff base dinucleating macrocycle led to intramolecular *m*-xylyl ring

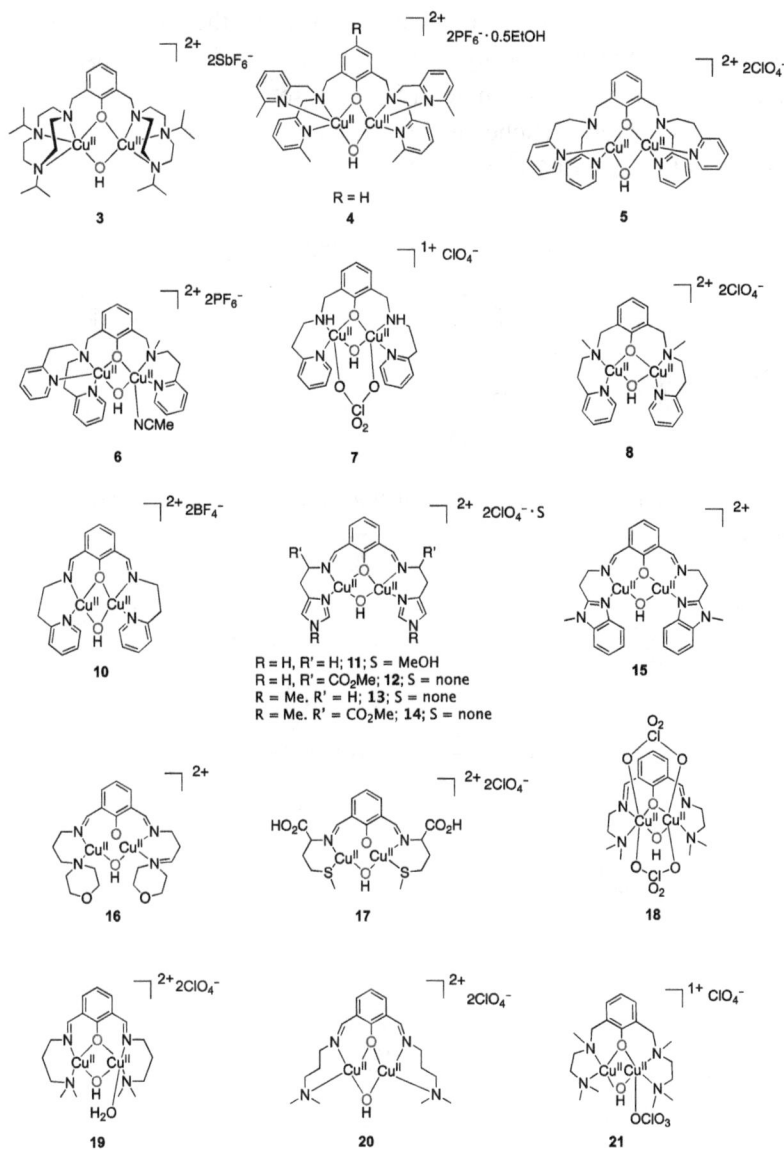

Figure 8. *m*-Xylyl ring hydroxylation model systems, using a large variety of *m*-xylyl-based ligands.

hydroxylation with concomitant hydrolysis of the macrocycle (**22**, Scheme 2).[25q] However, no hydrolysis of macrocycles was observed for similar reactions with 24-atom tetra Schiff base[25r,s] and 28-atom[25t,u] dinucleating macrocycles (Scheme 3); aromatic ring hydroxylation was observed in the case of **23**[25r,s] and **24**.[25t,u]

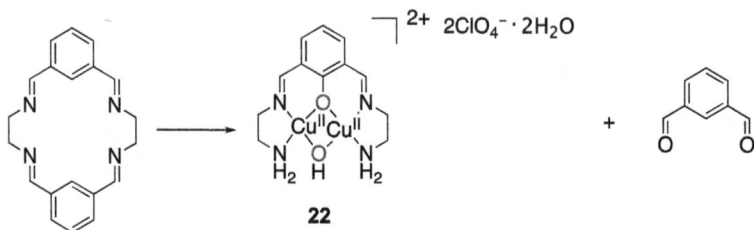

Scheme 2. Intramolecular ring hydroxylation with concomitant hydrolysis of the macrocycle.

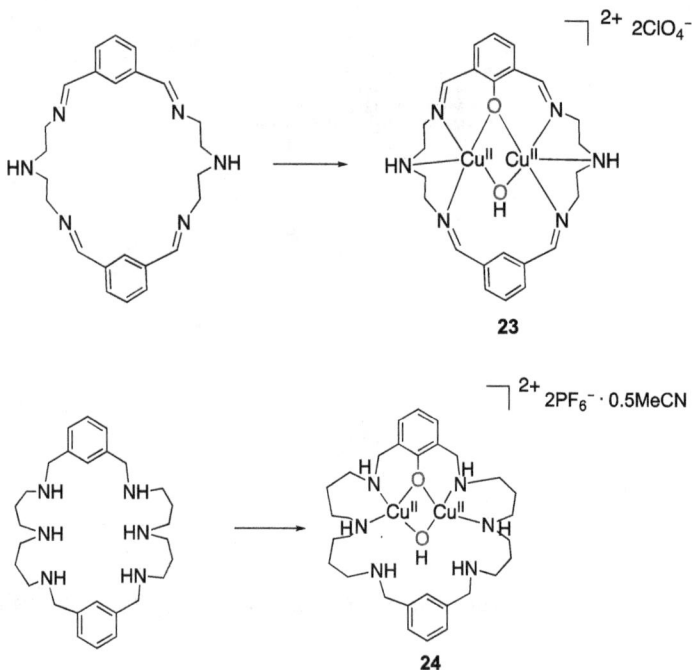

Scheme 3. Intramolecular ring hydroxylation with intact macrocyclic ligand systems.

C. Ring Hydroxylation Reactions with Non-*m*-Xylyl-Based Ligands

Now we discuss examples of aromatic ring hydroxylation reactions with non-*m*-xylyl-based open-chain ligands. The reactivity of $\{Cu_2O_2\}$ core was evidenced by Tolman and coworkers.[26a] The systems laid the foundations for a mechanistic understanding of the tyrosinase reaction. This reactivity was reinvestigated by Schindler, Tuczek, Holthausen, and coworkers.[26b] Thus, dinuclear copper complexes exhibiting bis-μ-oxido structure have been discovered to be able to mediate aromatic hydroxylation reactions in analogy to tyrosinase. The complexes **25** and **26** were obtained (Scheme 4). Although a bis-μ-oxido intermediate has not been observed in the enzyme as a stable intermediate, it may be relevant to the reactivity of tyrosinase. Stack and coworkers with simple diamine systems proposed an alternative aromatic hydroxylation mechanism.[27] Using a *m*-xylyl-derived macrocyclic ligand, the complex **27** was obtained.[28] Notably, using a tailor-made ligand with an appended phenol group, the formation of **28** was observed.[29] It is the first example provided by Tuczek and coworkers, mediating the monooxygenation of a phenol by the $\{Cu_2O_2\}$ core, in the absence of an external base.

D. Oxidation of External Substrates by Dicopper Systems

Initially, mainly aromatic *m*-xylyl ring hydroxylation reactions by $\{Cu_2P^S\}$ and $\{Cu_2O_2\}$ cores were investigated. Subsequently, the focus of interest shifted into hydroxylation of external mono- and diphenolic substrates, with an additional challenge to develop catalytic tyrosinase-model systems.

Using a *m*-xylyl-based ligand system (Scheme 1), Karlin and coworkers were the first to demonstrate spectroscopically at low temperature the formation of side-on peroxodicopper(II).[21] Casella and coworkers also showed the existence of $\{Cu_2P^S\}$ with **29** (R = Me).[30a] Costas, Casella, Rybak-Akimova, Que, and coworkers provided evidence for the formation of $\{Cu_2O_2\}$ **30**.[30b,c] Using an unsymmetrical ligand, Costas, Luis, Ribas, Que, Casella, and coworkers demonstrated the existence of an $\{Cu_2P^E\}$ core **31**.[30d] The formation of $\{Cu_2O_2\}$ **32**[30e] and $\{Cu_2P^E\}$ **33**[30d]

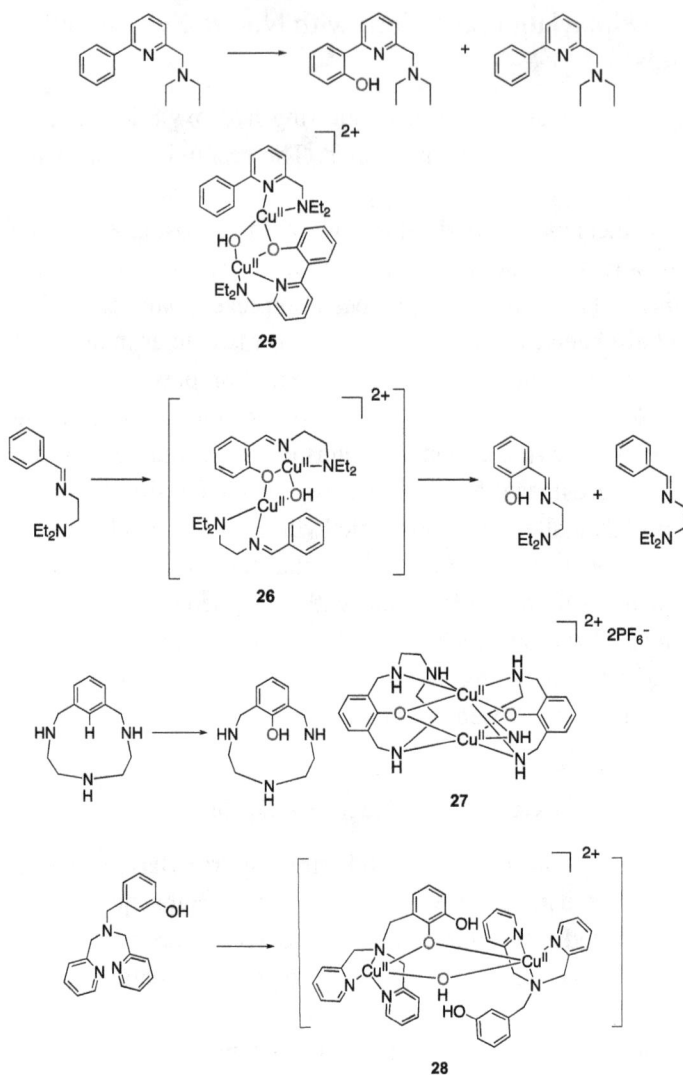

Scheme 4. Intramolecular ring hydroxylation with non-*m*-xylyl-based ligand systems.

were also reported with symmetrical Schiff base macrocyclic ligands. The intermediates **29–32** are schematically displayed in Figure 9. The cores **30** and **31** have been shown to mediate tyrosinase-like phenolate *ortho*-hydroxylation reactions (Figure 9). Apart from abovementioned intramolecular aromatic ring hydroxylation reactions, a couple of biomimetic

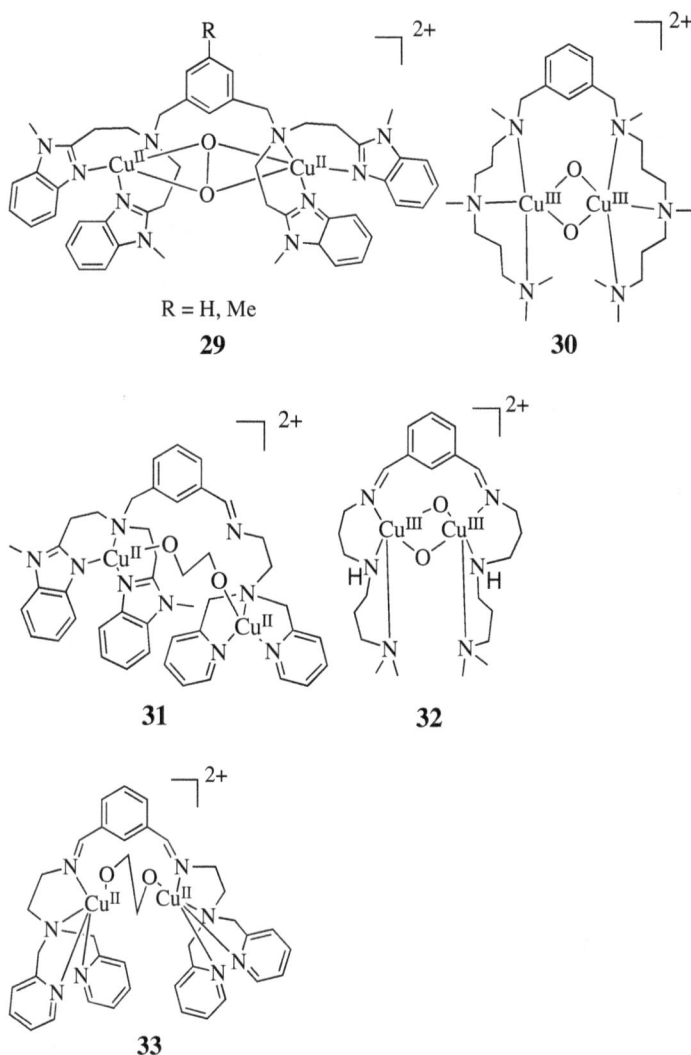

Figure 9.   Formation of {Cu_2P^S}, {Cu_2O_2}, {Cu_2P^E}, and {Cu_2O_2} cores.

binuclear copper complexes were synthesized and tested for their capability to convert external monophenolic substrates to *o*-diphenols or *o*-quinones.

Réglier *et al.* reported[31a] that a mixture of a ligand, containing pyridylethylimine sidearms bridged by a biphenyl spacer (Figure 10) placed

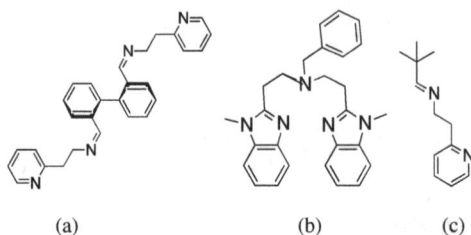

(a)　　　　　　　(b)　　　　(c)

Figure 10. (a) Réglier's ligand with biphenyl spacer, (b) tridentate analogue of the Casella's ligand used in **29** (Figure 9), and (c) bidentate analogue of Réglier's biphenyl spacer ligand.

with two equiv of Cu(I) salt, 100 equiv of 2,4-di-*tert*-butylphenol (DTBP–H), and 200 equiv of Et_3N led to the catalytic generation of quinone with a turnover number (TON) of 16 (Figure 10a). The progress of the reaction was monitored by the appearance of the optical absorption band of 3,5-di-*tert*-butyl-*o*-quinone (DTBQ) at ~400 nm. After 1h, the reaction stopped, presumably by the formation of an oxo-bridged dicopper(II) complex. A mechanism was proposed for this reaction, involving the formation of a $\{Cu_2P^S\}$ and a catecholate-bridged dicopper(II) intermediate. Stoichiometric hydroxylation of phenols to quinones has also been shown for a mononuclear Cu(I) complex supported by the tridentate analogue (Figure 10b) of the dinucleating ligand present in **29** (see Figure 9). In this reaction an involvement of the $\{Cu_2P^S\}$ intermediate was demonstrated.[31b]

Tuczek and coworkers reported the first catalytic tyrosinase model system, based on a mononuclear four-coordinate Cu(I) complex supported by a bidentate ligand (a modified version of Réglier ligand) (Figure 10c) and two MeCN.[31c] Oxygenation of a mixture of this Cu(I) complex with a phenolate (ArO⁻) and Et_3N with 50 equiv of DTBP–H and 100 equiv of Et_3N in CH_2Cl_2 was found to catalytically generate DTBQ, with a TON of 18. Oxygenation initially formed the $\{Cu_2P^S\}$ intermediate, containing two coordinated phenolates (Figure 11). In this complex only one of the bound phenolates was hydroxylated, leading to an asymmetrically coordinated μ-catecholato-μ-hydroxo-bridged intermediate **34** (Figure 11). In a stoichiometric mode it allowed the two consecutive stages of the tyrosinase reaction, phenol hydroxylation, and product release as quinone, to be addressed individually.

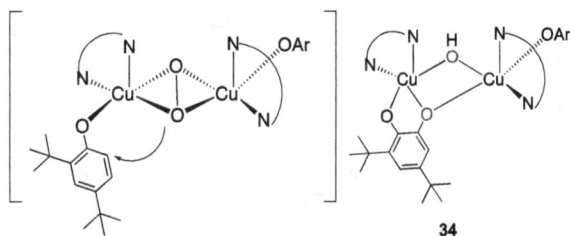

34

Figure 11. Intermediates (a) containing two coordinated phenolates and (b) asymmetrically coordinated μ-catecholato μ-hydroxo-bridging.

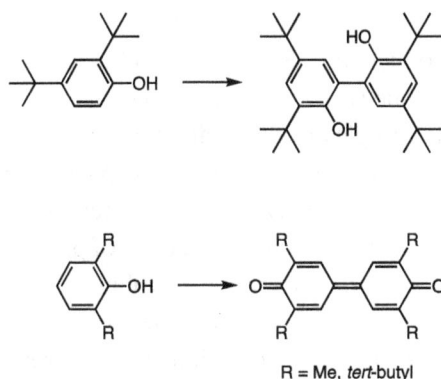

R = Me, *tert*-butyl

Scheme 5. Catalytic system forming bis-phenol and diphenoquinones, using F in the 2-position of the xylyl ring of the ligand used in 8 (Figure 8).

Oxygenation of a dicopper(I) complex of a *m*-xylyl-based ligand used by one of us in the synthesis of **8**,[25e,f] with a selective substitution of a fluorine atom in the 2-position of the ring, which underwent ring hydroxylation in **8**, proved to be a catalytic system in the promotion of oxidative carbon–carbon coupling of hindered phenols, which leads to the bis-phenol, 3,3′,5,5′-tetra-*tert*-butyl-2,2′-dihydroxybiphenyl and diphenoquinones, 3,3′,5,5′-tetra-*tert*-butyl-4,4′-diphenoquinone and 3,3′,5,5′-tetramethyl-4,47-diphenoquinone (Scheme 5). The reaction stopped with the formation of a dihydroxo-bridged dicopper(II) complex.[32] Unfortunately, we could not detect any copper–dioxygen intermediate.

Using 2,4-di-*tert*-butylphenolate as the substrate, the $\{Cu_2O_2\}$ intermediate generated (THF, 183 K) with the ligand present in **21** (Figure 8), we demonstrated[25p] the formation of C–C bonded bis-phenol in 42%

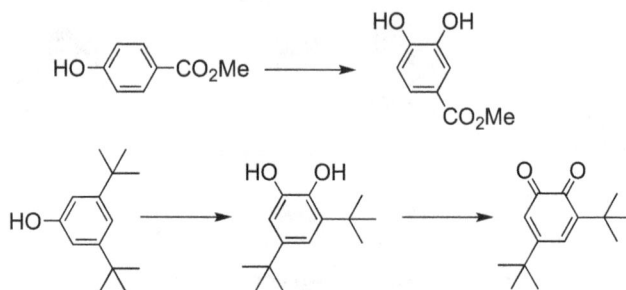

Figure 12. Tyrosinase-like chemical reactivity of **29** (R = H).

yield, catechol product in 38% yield (radical-coupled product; see Scheme 5), and quinone (Figure 12) in 8% yield.

The dicopper(I) complex of the ligand present in **29** (R = H; Figure 9) exhibits tyrosinase-like activity on exogenous phenols in the presence of dioxygen (Figure 12). With easily oxidizable phenols the reaction is catalytic but eventually leads to a complex mixture of products, as does tyrosinase (Figure 2). With more oxidation-resistant phenols the reaction is stoichiometric and stops at the level of catechol.[33a] The reaction suffers some limitation in the operation conditions in that a rigorously anhydrous and non-protic medium is required and formation of a phenolate adduct of the Cu(I) complex prior to the reaction with O_2 is also required, otherwise simple copper(I) oxidation occurs.[5e]

Experiments performed based on *in situ* generation of the phenolate showed that the dicopper(I) complexes of the following ligands (Scheme 6) in MeCN at 298 K exhibit tyrosinase-like monooxygenase activity in the presence of O_2. Yield of isolated catechol (~20–40%), in terms of chelate-ring size, follows the order: 6,6 > 5,6 > 5,5 > 5′,6 (5′ stands for five-membered chelate ring forming pyridine ring) of the ligand systems (Scheme 6). The catechol is the only product of phenol hydroxylation, when the reaction is carried out at 233 K.[33b]

It has been demonstrated that a {Cu_2O_2} core **30** (Figure 9) is competent to bind and hydroxylate phenolates (Figure 13b(i)).[30c] Exclusive formation of bis-oxido species is observed, before and after phenolate binding to the Cu_2O_2 site.[30c]

An asymmetric end-on peroxo-dicopper(II) complex **31** (Figure 9), with available coordination site, selectively binds phenolates and mediates

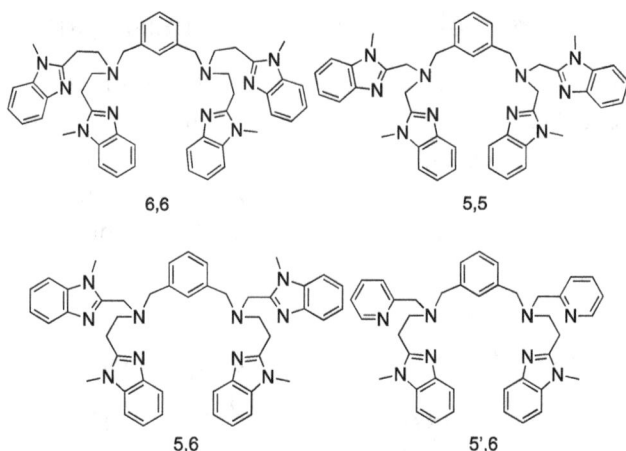

Scheme 6. Different dinucleating ligand systems for phenol hydroxylation.

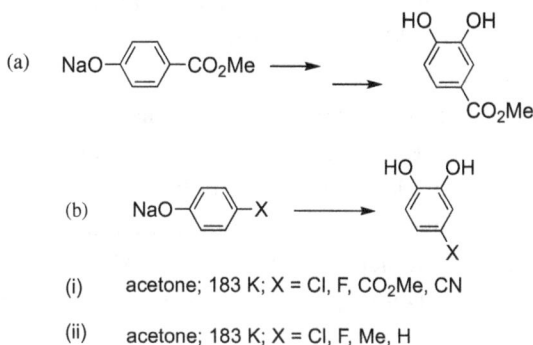

Figure 13. Tyrosinase-like activity by **29–31** intermediates: (a) with **29**, (b) with **30**, and (c) with **31**.

its *ortho*-hydroxylation (Figure 13b(ii)), thus functionally mimicking tyrosinase activity.[30d] This report was the first to demonstrate electrophilic character of the end-on $\{Cu_2P^E\}$ core, which was not observed for symmetric analogues.[30d] DFT computations were performed to assess the stability of different isomers ($\{Cu_2P^E\}$, $\{Cu_2P^S\}$, and $\{Cu_2O_2\}$), and it was found that the $\{Cu_2P^E\}$ is the most stable out of three species. So, the authors concluded that the $\{Cu_2P^E\}$ core is involved in aromatic hydroxylation reaction. Not many DFT studies are available on $\{Cu_2P^E\}$, compared to $\{Cu_2P^S\}$ and $\{Cu_2O_2\}$ cores (see below).

V. Theoretical Studies on Tyrosinase Models

Now we turn attention to the discussion on electronic structural properties from theoretical calculations. A brief discussion on the structure of *oxy*-Tyr and three Cu_2O_2 cores, identified and structurally and spectroscopically characterized, has already been done in Section **1.3**. In this section, we will discuss the frontier molecular orbitals (FMOs) of the three Cu_2O_2 cores.[8a]

A. Frontier Molecular Orbitals of the Cu_2O_2 Cores

1. $\mu\text{-}\eta^2{:}\eta^2$-*Peroxo-dicopper(II)*: $\{Cu^{II}_2(\mu\text{-}\eta^2{:}\eta^2 O_2)\}^{2+} \{Cu_2P^S\}$

During the $\{\mathbf{Cu_2P^S}\}$ core formation, the two Cu(I) centers interact with O_2 resulting in the oxidation of Cu(I) to Cu(II) and reduction of O_2 to $O_2{}^{2-}$. Each Cu(II) center would have an unpaired electron. The two unpaired spins of Cu could exist either in parallel (triplet) or anti-parallel (open-shell singlet) configuration. The $\{\mathbf{Cu_2P^S}\}$ cores found in tyrosinase and its biomimetic complexes are EPR silent, thus reflecting that the two unpaired spins of Cu are anti-parallel and strongly antiferromagnetically coupled. For a better understanding of the bonding picture in $\{\mathbf{Cu_2P^S}\}$, the FMOs are presented (Figure 14), in which half-filled $Cu(d_{x^2-y^2})$ orbitals are shown to interact with $O_2{}^{2-}$ valence MOs (in-plane $\pi_{ip}{}^*$, out-of-plane $\pi_{op}{}^*$, and σ^*). Here, $\pi_{ip}{}^*$ and $\pi_{op}{}^*$ are doubly occupied and σ^* is a vacant MO.

The FMO diagram of $\{\mathbf{Cu_2P^S}\}$ displays that the $\pi_{op}{}^*(O{-}O)$ orbital of $O_2{}^{2-}$ is perpendicular to the Cu_2O_2 plane, thus $\pi_{op}{}^*(O{-}O)$ brings a non-bonding interaction with Cu d-orbitals and forms HOMO-1. The $\pi_{ip}{}^*(O{-}O)$ orbital lying in the Cu_2O_2 plane overlaps with the $Cu(d_{x^2-y^2})$ orbitals and provides a bonding (HOMO-2) and an antibonding (LUMO) FMOs. The bridging $\pi_{ip}{}^*(O{-}O)$ orbital in HOMO-2 provides a super-exchange pathway to achieve strong antiferromagnetic coupling. Thus, tyrosinase and synthetic inorganic complexes containing $\{\mathbf{Cu_2P^S}\}$ cores are EPR-silent. The HOMO in $\{\mathbf{Cu_2P^S}\}$ is formed due to back-bonding from filled $Cu(d_{x^2-y^2})$ orbitals to the empty $\sigma^*(O{-}O)$ orbital. The back-donation from $Cu(d_{x^2-y^2})$ to $\sigma^*(O{-}O)$ weakens the $\sigma(O{-}O)$ bond of $\{\mathbf{Cu_2P^S}\}$, compared to a typical peroxide ion ($O_2{}^{2-}$).

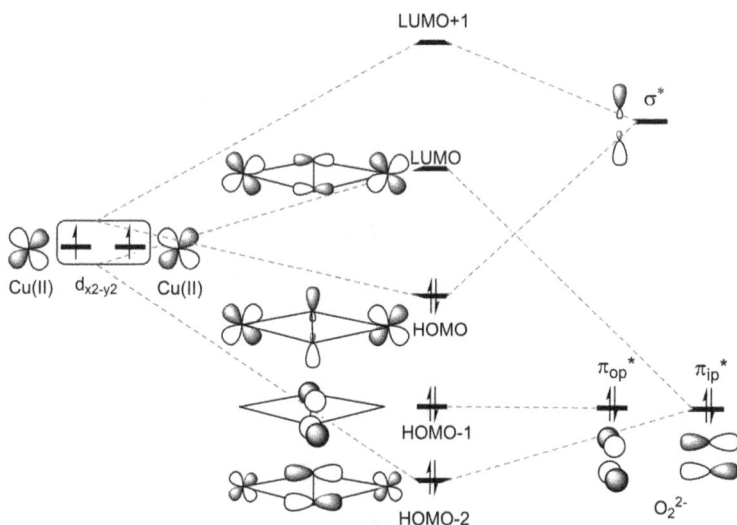

Figure 14. Frontier molecular orbitals of a $\{Cu^{II}_2(\mu\text{-}\eta^2\text{:}\eta^2\text{-}O_2)\}^{2+}$ core.

2. Trans-μ-1,2-peroxo-dicopper(II): $\{Cu^{II}_2(\mu\text{-}\eta^1\text{:}\eta^1\text{-}O_2)\}^{2+}$ $\{Cu_2P^E\}$

Similar to $\{Cu_2P^S\}$ (Section 1), $\{Cu_2P^E\}$ cores also contain two Cu(II) centers bound to a O_2^{2-} ligand; however, unlike a η^2-binding mode as in $\{Cu_2P^S\}$, $\{Cu_2P^E\}$ cores have a η^1-binding mode between the two Cu(II) and O_2^{2-} ions (Figure 5). In $\{Cu_2P^E\}$, the in-plane $\pi_{ip}*(O\text{-}O)$ orbital of O_2^{2-} overlaps with the $Cu(d_z^2)$ orbitals to form a bonding (HOMO-2) and an antibonding (LUMO) FMOs (Figure 15). The $Cu(d_z^2)$ orbitals attain a head-to-head overlap with $\pi_{ip}*(O\text{-}O)$. In $\{Cu_2P^S\}$ cores, the $Cu(d_x^2\text{-}y^2)$ and $\pi_{ip}*(O\text{-}O)$ orbitals do not have head-to-head overlap like in $\{Cu_2P^E\}$ cores. Thus, the Cu–O distances in $\{Cu_2P^S\}$ (~1.92 Å) are longer compared to $\{Cu_2P^E\}$ (~1.85 Å).

The O–O distance in $\{Cu_2P^S\}$ (~1.42 Å) and $\{Cu_2P^E\}$ (~1.43 Å) are comparable as in both the cores only the O_2 π-bond is broken and the σ-bond is still intact. The $\{Cu_2P^S\}$ core is more compact than $\{Cu_2P^E\}$ as the distance between the two Cu in $\{Cu_2P^S\}$ (3.51 Å) is smaller than in $\{Cu_2P^E\}$ (4.36 Å). Thus, $\{Cu_2P^E\}$ species are preferred in large-size ligands, where due to the steric crowding of the ligands the two Cu cannot achieve a shorter distance. The inorganic complexes containing $\{Cu_2P^E\}$

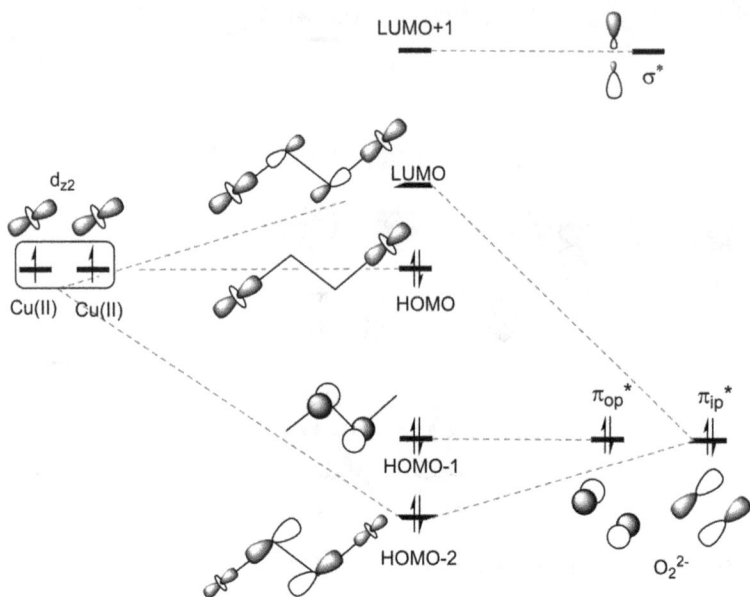

Figure 15. Frontier molecular orbitals of a $\{Cu_2(\mu\text{-}\eta^1\text{:}\eta^1\text{-}O_2)\}^{2+}$ core.

cores do not show EPR peaks, suggesting that the two Cu(II) centers are antiferromagnetically coupled. This antiferromagnetic interaction is achieved by following a super-exchange pathway in the bonding FMO (HOMO-2).

3. Bis(μ-oxido)dicopper(III): $\{Cu^{III}_2(\mu\text{-}O)_2\}^{2+}$ $\{Cu_2O_2\}$

A $\{Cu_2O_2\}$ core has two Cu(III) centers bound to two bridging O^{2-} ligands (Figure 5). The $\{Cu_2O_2\}$ cores can be derived from $\{Cu_2P^S\}$ by the oxidation of Cu(II) to Cu(III) and the reduction of O_2^{2-} to $2O^{2-}$ (oxides). The σ-bond of dioxygen in $\{Cu_2O_2\}$ is completely broken. Therefore, the distance between the two oxygens in $\{Cu_2O_2\}$ is longer (~2.3 Å) than the O–O distance in $\{Cu_2P^S\}/\{Cu_2P^E\}$ (~1.4 Å) and O_2 (~1.2 Å) (Figure 5).

The FMO diagram of $\{Cu_2O_2\}$ demonstrates the interaction of vacant $Cu(d_{x^2-y^2})$ orbitals with filled oxide orbitals (π_{ip}^*, π_{op}^*, and σ^*) (Figure 16). As for $\{Cu_2P^S\}$ and $\{Cu_2P^E\}$, the out-of-plane π_{op}^* orbital

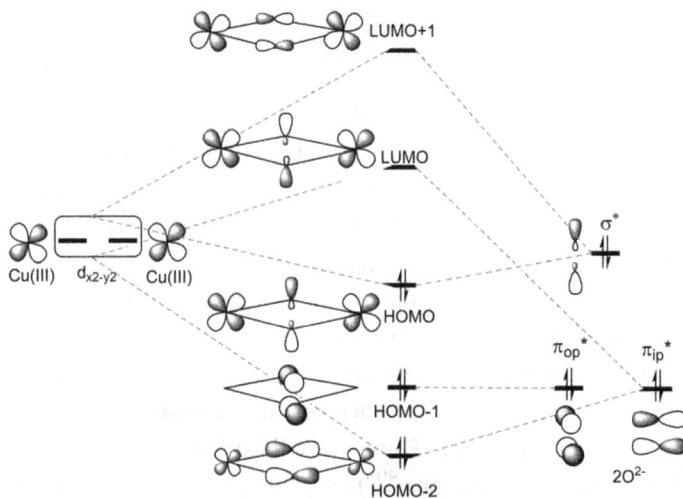

Figure 16.   Frontier molecular orbitals of a $\{Cu^{III}_2(\mu\text{-}O)_2\}^{2+}$ core.

and $Cu(d_{x^2-y^2})$ orbitals interact to form a non-bonding HOMO-1 FMO. The in-plane $\pi_{op}*$ orbital overlaps with $Cu(d_{x^2-y^2})$ orbitals to produce a bonding (HOMO-2) and an antibonding (LUMO+1) FMOs. Unlike $\{Cu_2P^S\}$ and $\{Cu_2P^E\}$, the $\sigma*$ orbital lies at a similar energy level as that of $Cu(d_{x^2-y^2})$ orbitals. Thus, an effective overlap between the two orbitals takes place resulting in HOMO and LUMO FMOs and a stable Cu–O covalent bond. The Cu–O distance in $\{Cu_2O_2\}$ is ~1.82 Å, which is shorter than the Cu–O distance in $\{Cu_2P^S\}$ (~1.85 Å)/$\{Cu_2P^E\}$ (~1.92 Å), indicating a strong σ(Cu–O) bond in $\{Cu_2O_2\}$. Due to elongation between the two oxygens in $\{Cu_2O_2\}$, the core has a compact motif (d(Cu···Cu) ~2.80 Å) than the other two isomers $\{Cu_2P^S\}$ (d(Cu···Cu) ~3.51 Å)/$\{Cu_2P^E\}$ (d(Cu···Cu) ~4.36 Å). So, sterically demanding bulky ligands do not support the formation of $\{Cu_2O_2\}$ motifs.

We discussed previously that $\{Cu_2O_2\}$ cores can be derived from $\{Cu_2P^S\}$ by pulling the electron density from two Cu(II) centers to $\sigma*$(O–O) orbital, thus breaking the dioxygen bond of O_2^{2-}. So, an interconversion between the two cores is possible. Indeed, such interconversions have been detected in synthetic laboratories[7b] and studied using various computational methods.

B. Interconversion Between {Cu_2P^S} and {Cu_2O_2} Cores: A Torture Track for Computations

Experimentally in 1996, Tolman and coworkers established that an interconversion between {Cu_2P^S} and {Cu_2O_2} cores is operative in solution.[34] It was found that this interconversion is sensitive to counter-anions, solvents, and temperature used in the reaction. Theoretically, such core transformations were studied in detail by Cramer,[35] Siegbahn,[36] Neese,[37] and Holthausen[38] groups. Based on their computational studies, these groups showed that the interconversion between the two isomers is not trivial for computational methods such as DFT. Cramer has named this conversion a "torture track" for theoretical chemists,[39] and even after 10 years of Cramer's findings, Herres-Pawlis still named the conversion a "torture track" for computations.[40] Thus, the interconversion has become an interesting problem for theoretical studies. Various benchmark studies have been performed to find out the best computational method for the relative stability of the two isomers. To benchmark DFT methods for such interconversions, scientists chose either experimental results or high-level computational methods as references.

Cramer and coworkers constructed {Cu_2P^S} and {Cu_2O_2} computational models supported by ammonia ligands for DFT benchmark studies.[35,39] *Ab initio* multi-reference configuration interaction (MRCI)[41] and renormalized coupled-cluster [CR-CC-(2,3)][42] methods were used as reference calculations to perform benchmark DFT studies. Cramer group found that GGA density functionals (such as BLYP)[43,44] within the closed-shell singlet approach were the best in agreement with the chosen reference calculations. However, the drawback of this study was the very small model system with NH_3 as terminal ligands.

Later, Siegbahn and coworkers used a larger model system for the computations and taken experimental relative stability data of {Cu_2P^S} and {Cu_2O_2} for the reference.[36] The chosen model systems ({Cu_2P^S} and {Cu_2O_2}) were supported by $^iPr_3$-TACN (1,4,7-tri-isopropyl-triazacyclononane) ligands. In their study the authors found that the B3LYP* density functional[45] (with 15% Fock exchange) within the broken-symmetry DFT approach works well for relative stability comparison of the two isomers. They also included dispersion and relativistic effects in the computations and their results were in good agreement with experiments.

Computational studies by the Cramer group have suggested to use the closed-shell singlet approach with GGA density functionals (0% Fock exchange), whereas Siegbahn found that B3LYP* (15% Fock exchange) functional works well but broken-symmetry solutions should be considered.[36] So, Neese and coworkers re-investigated this isomerization problem using $\{Cu_2P^S\}$ and $\{Cu_2O_2\}$ motifs, supported by ethylenediamine ligands. To generate an accurate reference data for the relative stability of the two isomers, the group used *ab initio* single reference local pair natural orbital coupled cluster (LPNO-CCSD)[46] theory. The authors found that the B3LYP (20% Fock exchange)[47] density functional within the closed-shell solution, including dispersion and relativistic effects, is the best method for studying this isomerization.

All the benchmark studies presented so far were based on either experimental references or high-level computational references for the relative stability of the two isomers. In 2017, Holthausen and coworkers chose both experimental and high-level computational methods as references for their benchmark studies.[38] The experimental relative stability data was taken from the work of Stack's group and the computational reference was generated from the CCSD(T) calculations. The authors showed that GGA functionals works best but it is also important to include closed-shell solutions for the two isomers to compare their relative stability. Moreover, DFT calculation must include dispersion, solvent, and relativistic corrections to get the best comparison with references. Moreover, Holthausen and coworkers also did a DFT benchmark study for C–H oxidation step and proposed that GGA functionals also work well for the hydroxylation steps. Hence, at present GGA functionals are the best choice for studying the interconversion between $\{Cu_2P^S\}$ and $\{Cu_2O_2\}$, and subsequent C–H oxidation step.

C. $\{Cu_2P^S\}$, $\{Cu_2P^E\}$, and $\{Cu_2O_2\}$ Motifs in C–H Oxidation: DFT Studies

Tyrosinase utilizes a $\{Cu_2P^S\}$ intermediate for performing aromatic C–H oxidation of tyrosine. However, in synthetic biomimetic studies interconversion between $\{Cu_2P^S\}$ and $\{Cu_2O_2\}$ cores were detected. Hence, out of the two, which core is active for C–H oxidation/hydroxylation is always an open question. Similarly, $\{Cu_2P^E\}$ could also be a potential candidate for performing C–H hydroxylation. In this regard, computational mechanistic studies have been very successful. Here, we will present DFT-based mechanistic studies on aromatic C–H hydroxylation via $\{Cu_2P^S\}$, $\{Cu_2P^E\}$, and $\{Cu_2O_2\}$ motifs. These models duplicate the reactivity of the tyrosinase enzyme.

1. Aromatic C–H hydroxylation via $\{Cu_2P^S\}$ motifs

Holthausen and Schindler groups carried out a combined experimental and computational work on a copper complex supported by a m-xylyl-based dinucleating ligand (Figure 17) (structural drawing **20**; Figure 8).[250] In this work, the authors were successful in performing intramolecular

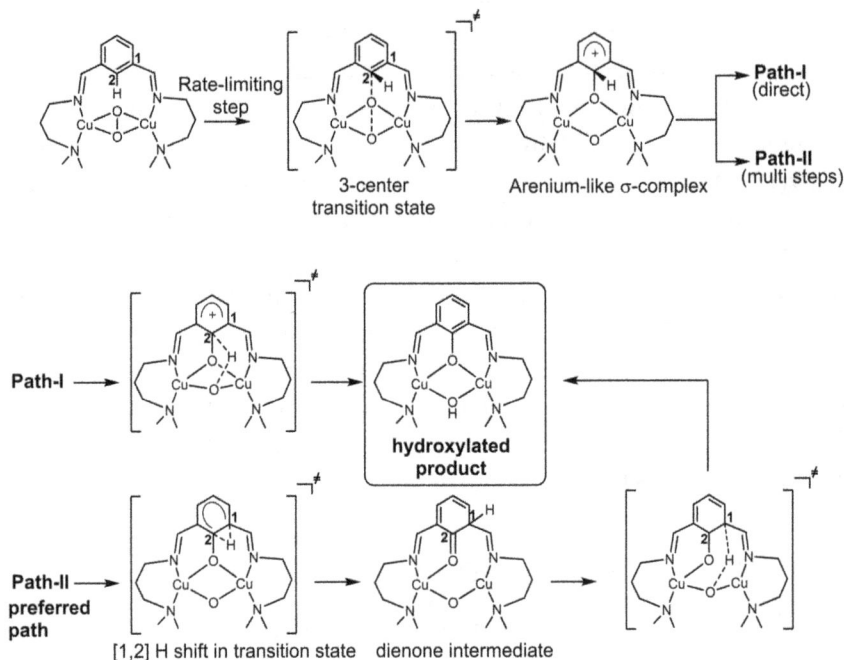

Figure 17. Intramolecular aromatic ligand hydroxylation via a $\{Cu_2P^S\}$ core.

C–H hydroxylation at the 2-position of the ligand. DFT studies revealed that the $\{Cu_2P^S\}$ core approaches toward 2-position of the *m*-xylene spacer and forms a 3-center transition state, in which O–O bond cleavage and C–O bond formation take place in a single step to produce an arenium-like σ complex. From this σ complex, two pathways leading to a single hydroxylated product were computed. Along Path-I, hydrogen present at the 2-position of the *m*-xylene ring is abstracted by the bridging μ-oxo atom of the Cu_2O_2 motif to generate the hydroxylated (μ-phenoxo μ-hydroxo) product in a single step. In Path-II, the σ complex converts into the hydroxylated product in two steps: first, the σ complex rearranges into a dienone intermediate *via* a 1,2-H shift across the aromatic ring. Such a hydrogen shift was earlier proposed by Karlin and denoted as the NIH shift.[48] In the second step, hydrogen present at the 1-position of the ring is abstracted by the bridging μ-oxo atom to yield the desired hydroxylated product. Out of the two mechanisms the second one, which proceeds *via* a dienone intermediate, is preferred based on the lower transition state barriers along the reaction steps. In this study, the authors showed that the $\{Cu_2P^S\}$ core is directly involved in the aromatic ligand hydroxylation, without the formation of a $\{Cu_2O_2\}$ intermediate. The 3-center transition state connected to $\{Cu_2P^S\}$ species and arenium-like σ complex is found to be the rate-limiting step.

In a more recent DFT-based mechanistic study by Hong and coworkers[49] used Karlin's tricoordinated dicopper(I) complex (Scheme 1)[21a,50] to study intramolecular ligand hydroxylation in presence of O_2 (Figure 18). Karlin's group observed *m*-xylyl ring hydroxylation at the 2-position. The group also studied the influence of *para*-substituent effect on the reaction rate. They concluded that an electron-donating group at para position increases the rate of hydroxylation, whereas electron-withdrawing groups decrease the reaction rate. To identify intermediate steps of this hydroxylation, Hong's group explored two mechanisms, which were earlier proposed by Holthausen and coworkers for aromatic ligand hydroxylation (structural type **19**) (Figure 17). They showed in their DFT study that Path-II (multi-step mechanism *via* a dienone intermediate) is preferred over Path-I (single-step direct hydrogen abstraction from bridged oxygen), based on lower transition state barriers along Path-II. The rate-limiting barrier is extracted from the very first step in which the $\{Cu_2P^S\}$

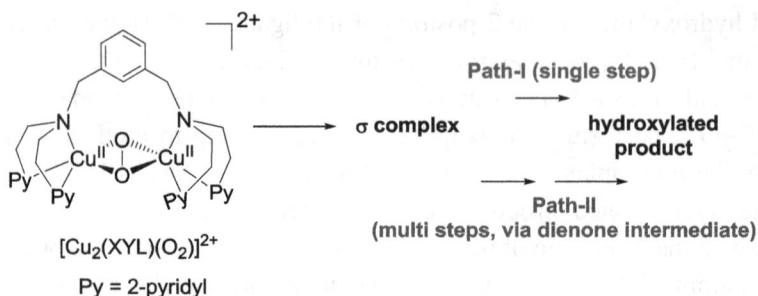

Figure 18. Intramolecular aromatic ring hydroxylation *via* a XYL-supported (Scheme 1) {**Cu₂Pˢ**} core. Detailed reaction paths are not shown, as they are like that shown in Figure 17.

motif attacks at the 2-position and builds a multi-center transition state. In the rate-determining transition step the O–O bond cleavage and C–O bond formation take place in a concerted fashion to yield an arenium-like σ complex. Like the Holthausen group, Hong and coworkers did not see a {**Cu₂O₂**}-type of species in their mechanistic investigations. The group also demonstrated theoretically the *p*-substituent effect by computing reaction barriers for the rate-determining step.

Based on DFT studies carried out by the Holthausen group and later by the Hong group, it can be concluded that dinuclear copper species supported by *m*-xylyl type ligands prefer aromatic ligand hydroxylation *via* dienone intermediate. However, such intermediate has not been detected in experimental findings. Their studies also suggest that {**Cu₂Pˢ**} is the active site and {**Cu₂O₂**} species does not exist in such cases.

Solà and coworkers used a large macrocyclic ligand-supported dicopper complex to study the intramolecular aromatic ligand hydroxylation (Figure 19).[25u] The said ligand contains two closely placed *m*-xylyl rings. In their DFT study they found that the two isomers ({**Cu₂Pˢ**} and {**Cu₂O₂**}) are not significantly different in energies. So, both cores are potential candidates for aromatic hydroxylation. The group studied intramolecular aromatic hydroxylation only *via* the {**Cu₂Pˢ**} intermediate; however, they also provide a note that the pathway *via* the {**Cu₂O₂**} core cannot be ruled out. Thus, in the initial step of the mechanism {**Cu₂Pˢ**} core attacks at the 2-position of the ring to yield a σ complex. In the next steps the other *m*-xylyl ring assists the hydrogen transfer from the

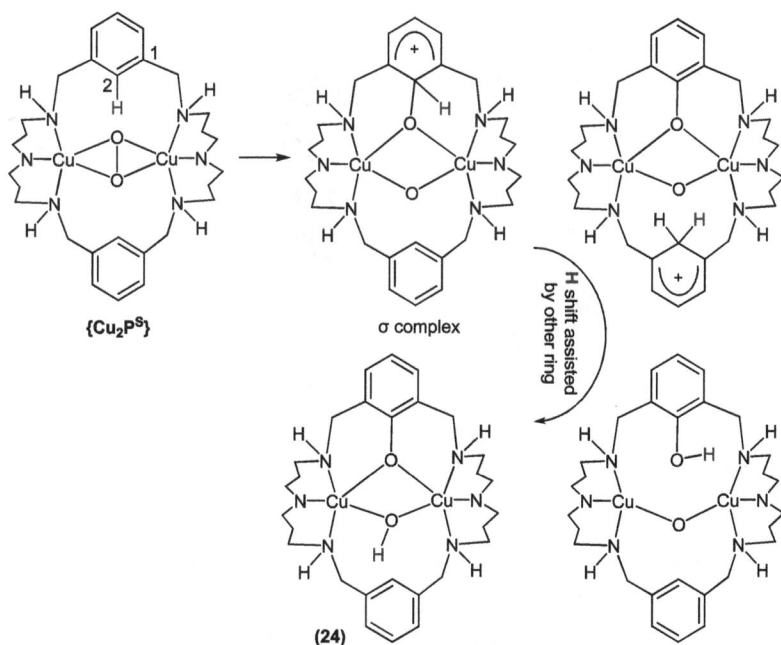

Figure 19. Intramolecular aromatic ligand hydroxylation *via* {Cu$_2$P^S} core to the product (**24**).

2-position of the first ring to the bridged oxygen and eventually yield a μ-phenoxo μ-hydroxo product. Hence, Solà and coworkers showed the use of a second *m*-xylyl ring for hydrogen transfer, such a mechanism could be envisioned in tyrosinase since histidine moieties close to tyrosine may assist the hydrogen transfer in aromatic C–H bond oxidation.

2. Aromatic C–H hydroxylation via {Cu$_2$O$_2$}

In this section, we discuss DFT studies on {Cu$_2$O$_2$}-mediated aromatic C–H hydroxylation reactions. Holthausen, Schindler, and Tuczek groups reported {Cu$_2$O$_2$}-mediated ligand-based aromatic hydroxylation.[26b] In this study these groups revisited Tolman's aromatic hydroxylation system[26a] and provided a detailed mechanistic study of this reaction. They also considered a new ligand in which the copper(I) complex on oxygenation-afforded salicylaldehyde due to ring hydroxylation/oxidation (**26**).

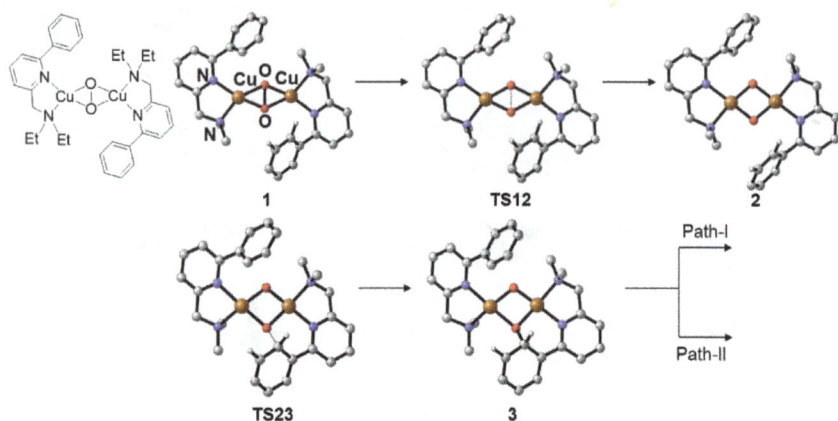

Figure 20. Formation of the σ-complex **3** from the {Cu$_2$O$_2$} intermediate in Tolman model.

DFT-based mechanism for Tolman's aromatic hydroxylation system is displayed in Figure 20.

DFT computations showed that in Tolman's system the {**Cu$_2$O$_2$**} intermediate **2** (it is not complex **2**) is significantly stable than its {**Cu$_2$P^S**} isomer **1** (Figure 20). Thus, the {**Cu$_2$O$_2$**} species would clearly be the dominant motif in the reaction mixture. The {**Cu$_2$O$_2$**} intermediate **2** attacks the nearest carbon of the phenyl ring to yield a σ-complex **3**. From the intermediate **3** (**1**, **2**, and **3** are not complex numbers) two pathways branch off.

Out of the two pathways the one that carries multiple steps and proceeds *via* a dienone intermediate is found to be the preferred one due to low activation barriers along the pathway (Figure 21). The rate-limiting transition state in the overall reaction is the 1,2-H shift (transition-state **TS34**), which results in a dienone intermediate **4**. Hence, the nature of the rate-limiting step is different in the {**Cu$_2$O$_2$**}-mediated aromatic hydroxylation than in {**Cu$_2$P^S**} hydroxylation (Figures 17, 18, and 20). In Path-I, **TS36** provides direct path to the hydroxylated product.

Holthausen and coworkers also computed theoretically the *p*-substituent effect on the rate of the reaction. They found that the groups CF$_3$, NO$_2$, CHO, and so on, which deactivate the benzene ring increase the barrier, whereas the groups OCH$_3$, OH, CH$_3$, and so on, which activate the

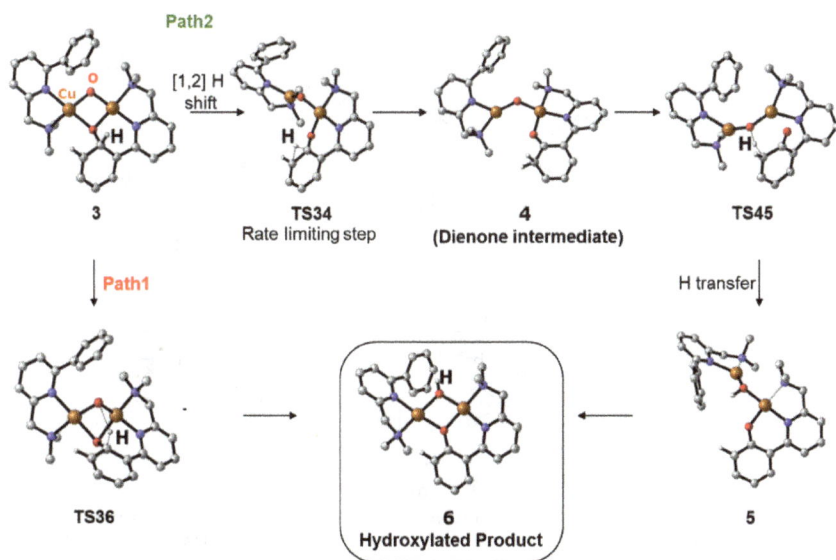

Figures 21. Two pathways leading to the hydroxylated product from the σ-complex 3 in the model complex of Tolman.

benzene ring decrease the barrier. Hence, the authors proposed that $\{Cu_2O_2\}$'s attack on the aromatic ring is an electrophilic substitution reaction.[26b]

The groups of Holthausen, Schindler, and Tuczek showed an intramolecular aromatic ligand hydroxylation *via* $\{Cu_2O_2\}$ species. Blomberg and coworkers employed DFT to investigate aromatic hydroxylation of an external phenolate *via* a $\{Cu_2O_2\}$ species (Figures 22).[51] For this purpose, the Blomberg group chose an experimental reaction established by the Stack group. In this reaction, Stack and coworkers used a DBED (DBED = *N,N'*-di-*tert*-butyl-ethylenediamine) supported copper/O_2 complex.[27] DFT-computed energetics of DBED-supported $\{Cu_2O_2\}$ and $\{Cu_2P^S\}$ systems confirmed that the two isomers are in equilibrium. However, after the addition of the external phenolate substrate in the reaction mixture, the equilibrium is shifted toward the phenolate-bound $\{Cu_2O_2\}$ isomer. In the next step, $\{Cu_2O_2\}$ core at attacks 2-position of the ring carbon (of phenolate) to yield a σ complex. Subsequently, the bridged μ-oxido abstracts H atom from 2-position of the ring to yield the hydroxylated

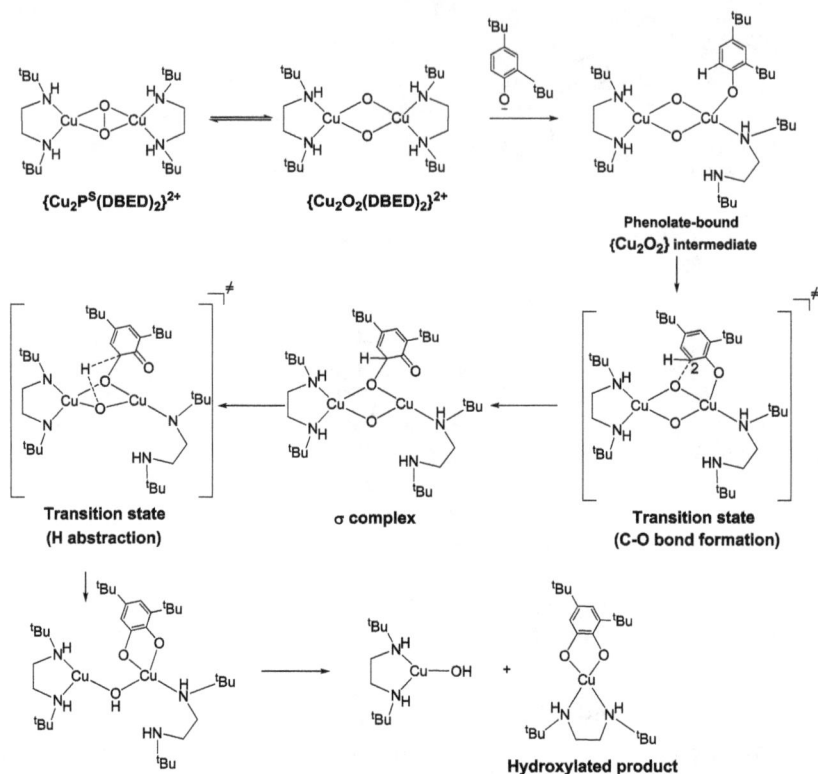

Figures 22. Hydroxylation of an external phenolate as substrate by a $\{Cu_2O_2\}$ core.

product. Thus, the Blomberg group showed in their computations that $\{Cu_2O_2\}$ cores are capable of hydroxylating aromatic C–H bond of external substrates.

In a recent study, Costas, Company, Luis, and coworkers utilized Stack's $\{Cu_2O_2(DBED)_2\}^{2+}$ system[27] to selectively activate a strong C–F bond of a phenolate substrate (Figure 23).[52] The authors also observed in their experimental studies that the *ortho*-hydroxylated product is minor whereas the *ortho*-defluorinated hydroxylated product is the major product. This finding was indeed non-intuitive as C–F bonds are stable than C–H bonds. So, activation of a C–F bond in the presence of a relatively weaker C–H bond was unclear (Figures 23). To gain insights of this finding the authors computed reaction pathways for the *ortho*-defluorinated hydroxylated product and the *ortho*-hydroxylated product.

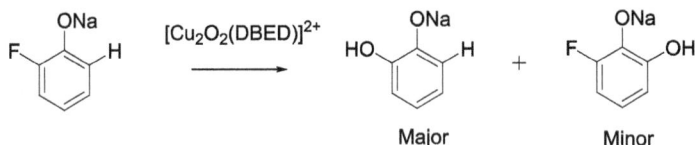

Figure 23. Selective C–F bond activation catalyzed by a {Cu₂O₂} core: {Cu^{III}₂O₂(DBED)₂}$^{2+}$.

Figure 24. Equilibrium between {Cu₂P^S} and {Cu₂O₂} cores (before and after phenolate binding).

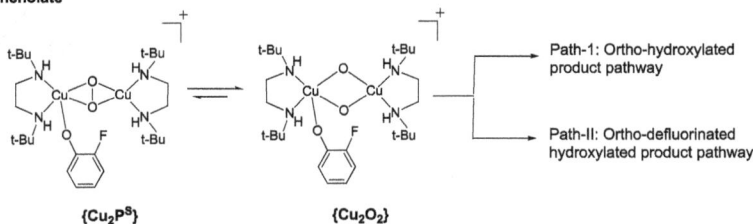

In their DFT calculations, these authors computed that the {Cu₂P^S} motif supported by DBED ligands is more stable than its corresponding {Cu₂O₂} motif. So, the equilibrium between the two cores would shift toward the {Cu₂P^S} species (Figure 24). However, 2-fluorophenolate bound {Cu₂O₂(DBED)₂}$^{2+}$ system is more stable for the {Cu₂O₂} core. Thus, after the addition of 2-fluorophenolate, {Cu₂O₂} species would dominate the reaction mixture. Similar results were also seen in Blomberg's DFT investigation (Figure 22). From here, two pathways branch off.

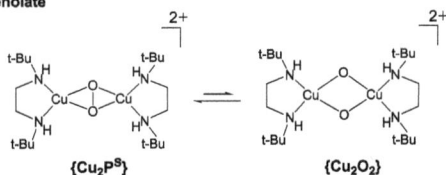

First pathway leads to the *ortho*-hydroxylated product (minor) and other provides a pathway to reach the *ortho*-defluorinated hydroxylated product (major) (Figure 25). The first pathway is like the one discussed in Figure 17. In Path-II, the phenolate substrate containing fluorine interacts with less-coordinated Cu(III). The reason for this interaction is the

Path-I

Path-II

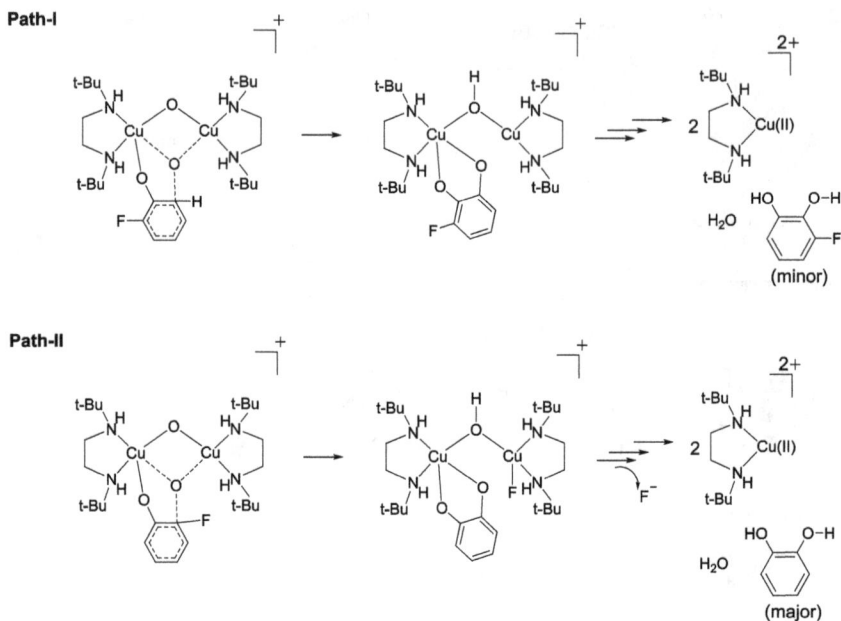

Figure 25. Pathways leading to *ortho*-hydroxylated product (minor in Path-I) and *ortho*-defluorinated hydroxylated product (major in Path-II).

attraction between a lone pair of F and Cu(III). Due to Cu···F interaction the C–F bond is found to be pre-activated. Such a pre-activation is not possible in ortho C–H hydroxylation as H does not contain lone pair. Thus, the attack of the $\{Cu_2O_2\}$ core at the carbon of the C–F bond is favorable over the carbon of the C–H bond. This results in a lower barrier for Path-II (Figure 25).

3. *Aromatic C–H hydroxylation via* $\{Cu_2P^E\}$

Garcia-Bosch *et al.* found that $\{Cu_2P^E\}$ cores are also capable of performing aromatic C–H hydroxylation (Figure 26; cf. Figure 9).[30d] They reacted the asymmetric $Cu^I_2(m\text{-xyl}^{N3N4})$ species with O_2[30d] and detected a $\{Cu_2P^E\}$ intermediate (structural drawing **31**). In the next step, sodium *p*-chlorophenolate was added in the reaction mixture with subsequent workup steps to yield *p*-chlorocatechol. DFT computations were performed on this reaction to compute the stability of different isomers

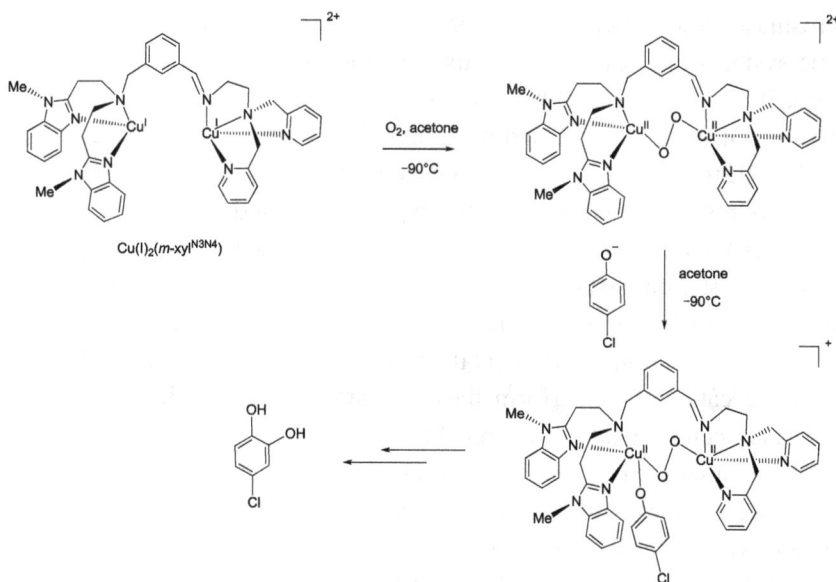

Figure 26. Aromatic ligand hydroxylation by a {Cu_2P^E} core.

({Cu_2P^E}, {Cu_2P^S}, {Cu_2O_2}), and it was found that the {Cu_2P^E} is the most stable species out of the three. So, the authors concluded that the {Cu_2P^E} core is involved in aromatic hydroxylation. Not many DFT studies are available on {Cu_2P^E} as compared to {Cu_2P^S} and {Cu_2O_2} cores.

VI. Conclusions

The use of dioxygen as an oxidant in organic transformations would very much benefit from fundamental knowledge on the mechanisms of its activation. Copper enzymes have been illustrated to process dioxygen in an efficient and purposeful manner. Modeling copper–dioxygen chemistry occurring in enzymes has attracted the attention of synthetic inorganic chemists. A powerful tool to extract mechanistic information on copper-catalyzed biochemical transformations, such as the tyrosinase reaction, is the study of small-molecule complexes, which mimic the active site structure of their biological counterparts. An additional challenge in copper–dioxygen chemistry has been the synthesis of catalytic model systems of tyrosinase. A significant challenge is the development of selective catalytic systems that use O_2 as the oxidant and avoid deleterious side reactions. Biological systems have unique capabilities in this regard through the control of the spatial and/or temporal distribution of substrates and oxidants. Such control has yet to be exerted in model systems and needs to be addressed, perhaps through more sophisticated biomimetic design strategies than those used so far. Further discoveries are likely to be made as mechanistic understanding of the enzymes accrues, new types of reactive intermediate in the biological and synthetic model systems are uncovered and applications toward catalysis are explored. Developing efficient synthetic catalysts to perform these transformations for industrial processes are central goals in chemical research.

In summary, this chapter briefly addresses the developments that have happened in the biomimetic front in benchmarking ligand design aspect to model tyrosinase-like activity and how theoretical calculations have strengthened our understanding from rationalization point of view. Copper-based oxidants and greater insights into biochemical oxidation processes with O_2 is a hot topic in bioinorganic chemistry of copper. From a humble beginning of intramolecular m-xylyl-based binucleating ligand hydroxylation first reported by Karlin and coworkers in early eighties, the biomimetic studies have enriched to advance our understanding of tyrosinase activity. The versatile reactivity properties of $\{Cu_2P^S\}$, $\{Cu_2O_2\}$, and $\{Cu_2P^E\}$ cores toward biorelevant substrates have been systematically investigated. The results have forced us to pay a new look at this area of research with sharpened direction.

VII. References

1. (a) Trammell, R.; Rajabimoghadam, K.; Garcia-Bosch, I. *Chem. Rev.* **2019**, *119*, 2954–3031. (b) Solomon, E. I.; Stahl, S. S. *Chem. Rev.* **2018**, *118*, 2299–2301.
2. Serrano-Plana, J.; Garcia-Bosch, I.; Company, A.; Costas, M. *Acc. Chem. Res.* **2015**, *48*, 2397–2406.
3. Que, L., Jr.; Tolman, W. B. *Nature* **2008**, *455*, 333–340.
4. (a) Solomon, E. I.; Heppner, D. E.; Johnston, E. M.; Ginsbach, J. W.; Cirera, J.; Qayyum, M.; Kieber-Emmons, M. T.; Kjaergaard, C. H.; Hadt, R. G.; Tian, L. *Chem. Rev.* **2014**, *114*, 3659–3853. (b) Solomon, E. I.; Chen, P.; Metz, M.; Lee, S.-K.;

Palmer, A. E. *Angew. Chem. Int. Ed.* **2001**, *40*, 4570–4590. (c) Solomon, E. I.; Sundaram, U. M.; Machonkin, T. E. *Chem. Rev.* **1996**, *96*, 2563–2605.

5. (a) Quist, D. A.; Diaz, D. E.; Liu, J. J.; Karlin, K. D. *J. Biol. Inorg. Chem.* **2017**, *22*, 253–288. (b) Garcia-Bosch, I.; Cowley, R. E.; Díaz, D. E.; Peterson, R. L.; Solomon, E. I.; Karlin, K. D. *J. Am. Chem. Soc.* **2017**, *139*, 3186–3195. (c) Hatcher, L. Q.; Karlin, K. D. *J. Biol. Inorg. Chem.* **2004**, *9*, 669–683. (d) Tyeklár, Z.; D. Karlin, K. D. *Acc. Chem. Res.* **1989**, *22*, 241–248. (e) Karlin, K. D.; Gultneh, Y. *Prog. Inorg. Chem.* **1987**, *35*, 219–327.

6. (a) Keown, W.; Gary, J. B.; Stack, T. D. P. *J. Biol. Inorg. Chem.* **2017**, *22*, 289–305. (b) Citek, C.; Herres-Pawlis, S.; Stack, T. D. P. *Acc. Chem. Res.* **2015**, *48*, 2424–2433. (c) Stack, T. D. P. *Dalton Trans.* **2003**, 1881–1889.

7. (a) Elwell, C. E.; Gagnon, N. L.; Neisen, B. D.; Dhar, D. Spaeth, A. D.; Yee, G. M.; Tolman, W. B. Chem. Rev. **2017**, *117*, 2059–2107. (b) Tolman, W. B. *J. Biol. Inorg. Chem.* **2006**, *11*, 261–271. (c) Holland, P. L.; Tolman, W. B. *Coord. Chem. Rev.* **1999**, 190–192, 855–869. (d) Tolman, W. B. *Acc. Chem. Res.* **1997**, *30*, 227–237.

8. (a) Mirica, L. M.; Ottenwaelder, X.; Stack, T. D. P. *Chem. Rev.* **2004**, *104*, 1013–1045. (b) Lewis, E. A.; Tolman, W. B. *Chem. Rev.* **2004**, *104*, 1047–1076.

9. (a) Schindler, S. *Eur. J. Inorg. Chem.* **2000**, 2311–2326. (b) Tolman, W. B. *Struct. Bonding (Berlin, Ger.)* **2000**, *97*, 179–211. (c) Kitajima, N.; Moro-oka, Y. *Chem. Rev.* **1994**, *94*, 737–757. (d) Kitajima, N. *Adv. Inorg. Chem.* **1992**, *39*, 1–77.

10. Itoh, S. Dicopper Enzymes. In *Comprehensive Coordination Chemistry – II: From Biology To Nanotechnology*; McCleverty, J. A.; Meyer, T. J., Eds. (Que Jr., L.; Tolman, W. B., Vol. Eds.); Elsevier/Pergamon: Amsterdam, 2004, Vol. 8, pp 369–393.

11. (a) Hatcher, L. Q.; Karlin, K. D. *Adv. Inorg. Chem.* **2006**, *58*, 131–184. (b) Karlin, K. D.; Tyeklár, Z. *Adv. Inorg. Biochem.* **1994**, *9*, 123–172.

12. (a) Itoh, S.; Fukuzumi, S. *Acc. Chem. Res.* **2007**, *40*, 592–600. (b) Itoh, S.; Tachi, Y. *Dalton Trans.* **2006**, 4531–4538. (c) Kunishita, A.; Osako, T.; Tachi, Y.; Teraoka, J.; Itoh, S. *Bull. Chem. Soc. Jpn.* **2006**, *79*, 1729–1741. (d) Komiyama, K.; Furutachi, H.; Nagatomo, S.; Hashimoto, A.; Hayashi, H.; Fujinami, S.; Suzuki, M.; Kitagawa, T. *Bull. Chem. Soc. Jpn.* **2004**, *77*, 59–72. (e) Itoh, S.; Fukuzumi, S. *Bull. Chem. Soc. Jpn.* **2002**, *75*, 2081–2095.

13. Kodera, M.; Kano, K. *Bull. Chem. Soc. Jpn.* **2007**, *80*, 662–676.

14. (a) Decker, H.; Schweikardt, T.; Tuczek, F. *Angew. Chem. Int. Ed.* **2006**, *45*, 4546–4550. (b) H. Decker, H.; Dillinger, R.; Tuczek, F. *Angew. Chem., Int. Ed.* **2000**, *39*, 1591–1595. (c) Decker, H.; Tuczek, F. *Trends Biochem. Sci.* **2000**, *25*, 392–397.

15. Battaini, G.; Granata, A.; Monzani, E.; Gullotti, M.; Casella, L. *Adv. Inorg. Chem.* **2006**, *58*, 185–233.

16. Que, Jr., L.; Tolman, W. B. *Angew. Chem. Int. Ed.* **2002**, *41*, 1114–1137.

17. (a) Solomon, E. I.; Lowery, M. D. *Science* **1993**, *259*, 1575–1581. (b) Karlin, K. D. *Science* **1993**, *261*, 701–708. (c) Ibers, J. A.; Holm, R. H. *Science* **1980**, *209*, 223–235.

18. (a) Magnus, K. A.; Hazes, B.; Ton-That, H.; Bonaventura, C.; Bonaventura, J.; Hol, W. G. J. *Proteins: Structure, Function and Genetics* **1994**, *19*, 302–309. (b) Hazes, B.; Magnus, K. A.; Bonaventura, C.; Bonaventure, J.; Dauter, Z.; Kalk, K. H.; Hol. W. G. J. *Protein Sci.* **1993**, *2*, 597–619.

19. (a) Matoba, Y.; Kumagai, T.; Yamamoto, A.; Yoshitsu, H.; Sugiyama, M. *J. Biol. Chem.* **2006**, *281*, 8981–8990. (b) Pillaiyar, T.; Manickam, M.; Namasivayam, V. *J. Enz. Inhib. Med. Chem.* **2007**, *32*, 403–425. (c) Parvez, S.; Kang, M.; Chung, H.-S.; Bae, H. *Phytother. Res.* **2007**, *21*, 805–816. (d) Zolghadri, S.; Bahrami, A.; Khan, M. T. H.; Munoz-Munoz, J.; Garcia-Molina, F.; Garcia-Canovas, F.; Saboury, A. A. *J. Enz. Inhib. Med. Chem.* **2019**, *34*, 279–309. (e) Shi, F.; Xie, L.; Lin, O.; Tong, C.; Fu, Q.; Xu, J.; Xiao, J.; Shi, S. *Food Chem.* **2020**, *312*, 126042.

20. (a) Klabunde, T.; Eicken, C.; Sacchettini, J. C.; Krebs, B. *Nat. Struct. Biol.* **1998**, *5*, 1084-1090. (b) Gerdemann, C.; Eicken, C.; Krebs, B. *Acc. Chem. Res.* **2002**, 35, 183–191. (b) Dey, S. K.; Mukherjee, A. *Coord. Chem. Rev.* **2016**, *310*, 80–115. (c) Ackermann, J.; Buchler, S.; Meyer, F. *C. R. Chimie* **2007**, *10*, 421–432. (d) Koval, I. A.; Gamez, P.; Belle, C.; Selmeczi, K.; Reedijk, J. *Chem. Soc. Rev.* **2006**, *35*, 814–840. (e) Selmeczi, K.; Réglier, M.; Giorgi, M.; Speier, G. *Coord. Chem. Rev.* **2003**, *245*, 191–201.

21. (a) Karlin, K. D.; Hayes, J. C.; Gultneh, Y.; Cruise, R. W.; McKown, J. W.; Hutchinson, J. H.; Zubieta, J. *J. Am. Chem. Soc.* **1984**, *106*, 2121–2128. (b) Karlin, K. D.; Nasir, M. S.; Cohen, B. I.; Cruse, R. W.; Kaderii, S.; Andreas D. Zuberbühler, A. D. *J. Am. Chem. Soc.* **1994**, *116*, 1324–1336. (c) Pidcock, E.; Obias, H. V.; Zhang, C. X.; Karlin, K. D.; Solomon, E. I. *J. Am. Chem. Soc.* **1998**, *120*, 7841–7847.

22. (a) De, A.; Mandal, S.; Mukherjee, R. *J. Inorg. Biochem.* **2008**, *102*, 1170–1189. (b) Spodine, E.; Manzur, J. *Coord. Chem. Rev.* **1992**, *119*, 171–198. (c) Sorrell, T.N. *Tetrahedron* **1989**, *45*, 3–68.

23. Mukherjee, R. Copper. In *Comprehensive Coordination Chemistry II: From Biology to Nanotechnology*; McCleverty, J. A.; Meyer, T. J., Eds. (Fenton, D. E., Vol. Ed.); Elsevier/Pergamon: Amsterdam, 2004; Vol. 6, pp 747–910.

24. Rolff, M.; Schottenheim, J.; Decker, H.; Tuczek, F. *Chem. Soc. Rev.* **2011**, *40*, 4077–4098.

25. (a) Mahapatra, S.; Kaderli, S.; Llobet, A.; Neuhold, Y.-M.; Palanché, T.; Halfen, J. A.; Young, V. G., Jr.; Kaden, T. A.; Que, L., Jr.; Zuberbühler, A. D.; Tolman, W. B. *Inorg. Chem.* **1997**, *36*, 6343–6356. (b) Matsumoto, T,; Furutachi, H.; Kobino, M.; Tomii, M.; Nagatomo, S.; Tosha, T.; Osako, T.; Fujinami, S.; Itoh, S.; Kitagawa, T.; Suzuki, M. *J. Am. Chem. Soc.* **2006**, *128*, 3874–3875. (c) Foxon, S. P.; Utz, D.; Astner, J.; Schindler, S.; Thaler, F.; Heinemann, F. W.; Liehr, G.; Mukherjee, J.; Balamurugan, V.; Ghosh, D.; Mukherjee, R. *Dalton Trans.* **2004**, 2321–2328. (d) Murthy, N. N.; Mahroof-Tahir, M.; Karlin, K. D. *Inorg. Chem.* **2001**, *40*, 628–635. (e) Ghosh, D.; Lal, T. K.; Ghosh, S.; Mukherjee, R. *Chem. Commun.* **1996**, 13–14. (f) Ghosh, D.; Mukherjee, R. *Inorg. Chem.* **1998**, *37*, 6597–6605. (g) Gelling, O. J.; van Bolhuis, F.; Meetsma, A.; Feringa, B. L. *Chem. Commun.* **1988**, 552–554. (h)

Casella, L.; Rigoni, L. *J. Chem. Soc., Comm. Commun.* **1985**, 1668–1669. (i) Casella, L.; Gullotti, M.; Pallanza, G.; Rigoni, L. *J. Am. Chem. Soc.* **1988**, *110*, 4221-4227. (j) Casella, L.; Gullotti, M.; Bartosek, M.; Pallanza, G.; Laurenti, E. *J. Chem. Soc., Comm. Commun.* **1991**, 1235–1237. (k) Gloria Alzuet, G.; Casella, L.; Villa, M. L.; Carugo, O.; Gullotti, M. *J. Chem. Soc., Dalton Trans.* **1997**, 4789–4794. (l) Sorrell, T. N.; Garrity, M. L. *Inorg. Chem.* **1991**, *30*, 210–215. (m) Mandal, S.; Mukherjee, R. *Inorg. Chim. Acta* **2006**, *359*, 4019–4026. (n) Drew, M. G. B.; Trocha-Grimshaw, J.; McKillop, K. P. *Polyhedron* **1989**, *8*, 2513–2515. (o) Sander, O.; Henß, A.; Näther, C.; Würtele, C.; Holthausen, M. C.; Schindler, S.; Tuczek, F. *Chem. Eur. J.* **2008**, *14*, 9714–9729. (p) Mandal, S.; Mukherjee, J.; Lloret, F.; Mukherjee, R. *Inorg. Chem.* **2012**, *51*, 13148–13161. (q) Gupta, R.; Mukherjee, R. *Inorg. Chim. Acta* **1997**, *263*, 133–137. (r) Menif, R.; Mattell, A. E. *J. Chem. Soc., Comm. Commun.* **1989**, 1521–1523. (s) Menif, R.; Martell, A. E.; Squattrito, P. J.; Clearfield, A. *inorg. Chem.* **1990**, *29*, 4723-4729. (t) Costas, M.; Ribas, X.; Poater, A.; Valbuena, J. M. L.; Xifra, R.; Company, A.; Duran, M.; Sola, M.; Llobet, A.; Corbella, M.; Usón, M. A.; Mahía, J.; Solans, X.; Shan, X.; Benet-Buchholz, J. *Inorg. Chem.* **2006**, *45*, 3569–3581. (u) Poater, A.; Ribas, X.; Llobet, A.; Cavallo, L.; Solà, M. *J. Am. Chem. Soc.* **2008**, *130*, 17710–17717.

26. (a) Patrick L. Holland, P. L.; Rodgers, K. R.; Tolman, W. B. *Angew. Chem. Int. Ed.* **1999**, *38*, 1139–1142. (b) Becker, J.; Gupta, P.; Angersbach, F.; Tuczek, F.; Näther, C.; Holthausen, M. C.; Schindler, S. *Chem. Eur. J.* **2015**, *21*, 11735–11744.

27. Mirica, L. M.; Vance, M.; Rudd, D. J.; Hedman, B.; Hodgson, K. O.; Solomon, E. I.; Stack, T. D. P. Science 2005, 308, 1890–1892.

28. Fusi, V.; Llobet, A.; Mahía, J.; Micheloni, M.; Paoli, P.; Ribas, X.; Rossi, P. Eur. J. Inorg. Chem. 2002, 987–990.

29. Hamann, J. N.; Rolff, M.; Tuczek, F. *Dalton Trans.* **2015**, *44*, 3251–3258.

30. (a) Palavicini, S.; Granata, A.; Monzani, E.; Casella, L. *J. Am. Chem. Soc.* **2005**, *127*, 18031–18036. (b) Company, A. ; Lamata, D.; Poater, A.; Solà, M.; Rybak-Akimova, E. V.; Que, Jr., L.; Fontodrona, X.; Parella, T; Llobet, A.; Costas, M. *Inorg. Chem.* **2006**, *45*, 5239–5241. (c) Company, A.; Palavicini, S.; Garcia-Bosch, I.; Mas-Ballesté, R.; Que, Jr., L.; Rybak-Akimova, E. V.; Casella, L.; Ribas, X.; Costas, M. *Chem. Eur. J.* **2008**, *14*, 3535–3538. (d) Garcia-Bosch, I.; Company, A.; Frisch, J. R.; Torrent-Sucarrat, M.; Cardellach, M.; Gamba, I.; Mireia Güell, M.; Casella, L.; Que, Jr., L.; Ribas, X.; Luis, J. M.; Costas, M. *Angew. Chem. Int. Ed.* **2010**, *49*, 2406–2409. (e) Company, A.; Gómez, L.; Mas-Ballesté, R.; Korendovych, I. V.; Ribas, X.; Poater, A.; Parella, T. ; Fontodrona, X.; Benet-Buchholz, J.; Solà, M.; Que, Jr., L.; Rybak-Akimova, E. V.; Costas, M. *Inorg. Chem.* **2007**, *46*, 4997–5012.

31. (a) Réglier, M.; Jorand, C.; Waegell, B. *J. Chem. Soc. Chem. Commun.* **1990**, 1752–1755. (b) Battaini, G.; De Carolis, M.; Monzani, E.; Tuczek, F.; Casella, L. *Chem. Commun.* **2003**, 726–727. (c) Rolff, M.; Schottenheim, J.; Peters, G.; Tuczek, F. *Angew. Chem. Int. Ed.* **2010**, *49*, 6438–6442.

32. Gupta, R.; Mukherjee, R. *Tet. Lett.* **2000**, *41*, 7763–7767.

33. (a) Casella, L.; Gullotti, M.; Radaelli, R.; Di Gennaro, P. *J. Chem. Soc., Chem. Commun.* **1991**, 1611–1612. (b) Casella, L.; Monzani, E.; Gullotti, M.; Cavagnino, D.; Cerina, G.; Santagostini, L.; Ugo, R. *Inorg. Chem.* **1996**, *35*, 7516–7525.

34. Halfen, J. A.; Mahapatra, S.; Wilkinson, E. C. ; Kaderli, S.; Young, V. G.; Que, L.; Zuberbühler, A. D.; Tolman, W. B. Science 1996, 271, 1397–1400.

35. Gherman, B. F.; Cramer, C. J. Coord. Chem. Rev. 2009, 253, 723–753.

36. Siegbahn, P. E. M.; Blomberg, M. R. A.; Chen, S.-L. J. Chem. Theory Comput. 2010, 6, 2040–2044.

37. Liakos, D. G.; Neese, F. J. Chem. Theory Comput. 2011, 7, 1511–1523.

38. Gupta, P.; Diefenbach, M.; Holthausen, M. C.; Förster, M.; Chem. Eur. J. 2017, 23, 1427–1435.

39. Cramer, C. J.; Wloch, M.; Piecuch, P.; Puzzarini, C.; Gagliardi, L. J. Phys. Chem. A 2006, 110, 1991–2004.

40. Rohrmüller, M.; Hoffmann, A.; Thierfelder, C.; Herres-Pawlis, S.; Schmidt, W. G. *J. Compt. Chem.* **2015**, *36*, 1672–1685.

41. (a) Werner, H.-J.; Knowles, P. J. *J. Chem. Phys.* **1988**, *89*, 5803–5814; (b) Knowles, P. J.; Werner, H. J. *Chem. Phys. Lett.* **1988**, *145*, 514–522.

42. (a) Kowalski, K.; Piecuch, P. *J. Chem. Phys.* **2000**, *113*, 18–35. (b) Piecuch, P.; Kowalski, K.; Pimienta, I. S. O.; Fan, P. D.; Lodriguito, M.; McGuire, M. J.; Kucharski, S. A.; Kuś, T.; Musiał, M. *Theor. Chem. Acc.* **2004**, *112*, 349–393. (c) Piecuch, P.; Wloch, M.; Gour, J. R.; Kinal, A. *Chem. Phys. Lett.* **2006**, *418*, 467–474.

43. (a) Lee, C.; Yang, W.; Parr, R. G. *Phys. Rev. B* **1988**, *37*, 785–789. (b) Becke, A. D. *Phys. Rev. A* **1988**, *38*, 3098–3100.

44. Miehlich, B.; Savin, A.; Stoll, H.; Preuss, H. Chem. Phys. Lett. 1989, 157, 200–206.

45. Reiher, M.; Salomon, O.; Artur Hess, B. *Theor. Chem. Acc.* **2001**, *107*, 48–55.

46. (a) Neese, F.; Hansen, A.; Liakos, D. G. *J. Chem. Phys.* **2009**, *131*, 064103–064117. (b) Liakos, D. G.; Hansen, A.; Neese, F. *J. Chem. Theory Comput.* **2010**, *7*, 76–87.

47. (a) Stephens, P. J.; Devlin, F. J.; Chabalowski, C. F.; Frisch, M. J. *J. Phys. Chem.* **1994**, *98*, 11623–11627. (b) Becke, A. D. *J. Chem. Phys.* **1993**, *98*, 5648–5652.

48. Nasir, M. S.; Cohen, B. I.; Karlin, K. D. J. Am. Chem. Soc. 1992, 114, 2482–2494.

49. Liu, Y. F.; Shen, J.; Chen, S.-L.; Qiao, W.; Zhou, S.; Hong, K. *Dalton Trans.* **2019**, *48*, 16882–16893.

50. (a) Karlin, K. D.; Dahlstrom, P. L.; Cozzette, S. N.; Scensny, P. M.; Zubieta, J. *J. Chem. Soc. Chem. Commun.* **1981**, 881–882. (b) Karlin, K. D.; Gultneh, Y.; Hutchinson, J. P.; Zubieta, J. *J. Am. Chem. Soc.* **1982**, *104*, 5240–5242.

51. Liu, Y. F.; Yu, J. G.; Siegbahn, P. E. M.; Blomberg, M. R. A. *Chem. Eur. J.* **2013**, *19*, 1942–1954.

52. Besalú-Sala, P.; Magallon, C.; Costas, M.; Company, A.; Luis, J. M. *Inorg. Chem.* **2020**, *59*, 17018–17027.

4 Monooxygenation of Phenols by Small-molecule Models of Tyrosinase: Correlations Between Structure and Catalytic Activity

**Alexander Koch*, Tobias A. Engesser*,
Ramona Jurgeleit*, and Felix Tuczek*,†**

*Institut für Anorganische Chemie, Christian-Albrechts-Universität
 zu Kiel, Max-Eyth Straße 2, 24118 Kiel, Germany
†ftuczek@ac.uni-kiel.de

I. Introduction

Copper enzymes are involved in important biochemical processes in animals or plants such as dioxygen transport and metabolism or electron transfer and are present in the three domains of life.[1,2] They are often categorized by classes (type I–IV, Cu_A, Cu_B, Cu_Z, and Cu_0), which derive from their structural environment or the spectroscopic properties of the metal center(s).[1,3] Type III refers to a (oxygen-binding) binuclear copper protein family comprising tyrosinase (Ty), tyrosinase-related proteins, catechol oxidase (CO), and hemocyanin.[1,3–5]

Tyrosinase is a phenoloxidase and catalyzes the conversion and oxidation of mono- and diphenols into the corresponding *ortho*-quinones whereas the catechol oxidase only exhibits the latter reactivity.[1,6–8] These enzymes mediate

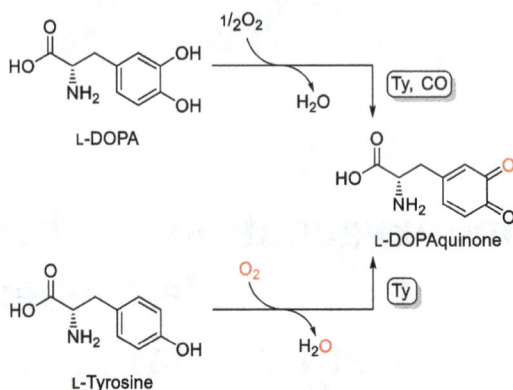

Scheme 1. Overview of the reactivities of tyrosinase (Ty) and catechol oxidase (CO).[6–8]

the oxygenation of L-tyrosine and, respectively, oxidation of L-DOPA to L-dopaquinone, which in turn polymerizes to melanin (Scheme 1).

II. Model Systems of Tyrosinase

For a better understanding of the reactivity and the functional differences of these enzymes and in addition to insights already obtained by structural and molecular biology, mechanistic investigations on small-molecule models can be beneficial.[7,9–11] In full analogy with their biological counterparts, the synthetic model systems of phenoloxidases can be divided into those exhibiting catechol oxidase and those exhibiting tyrosinase activities.[3,12–17] A further important distinction relates to the question of whether these systems are catalytic or not. Whereas the former in most cases applies to catechol oxidase models, most of the existing tyrosinase models perform monooxygenation reactions of aromatic substrates in a stoichiometric fashion.[7] By contrast, the number of catalytic model systems of tyrosinase is still very limited.[6,7,18]

An important factor in the investigation of copper-catalyzed oxidation and oxygenation reactions relates to the presence or absence of base. Whereas catechol oxidase reactions occur in the absence of base, monooxygenation reactions on external phenolic substrates in general require the addition of base to deprotonate phenolic substrates.[8]

In order to explore this correlation in more detail, several studies on inorganic model systems, which catalyze the monooxygenation of phenolic substrates, have been published in the past. This has provided insight into relevant steps of the monophenolase cycle, that is, the formation of the peroxo complex, the interaction of the substrate with the peroxo complex leading to a coordinated catecholate (including the role of base in this reaction), the two-electron oxidation of the bound catecholate and the release of the *ortho*-quinone product. As this chapter focuses on tyrosinase activity, small-molecule models of catechol oxidase will not be discussed here. Excellent reviews on the latter topic can be found elsewhere.[3,12-16]

A. Model Systems Performing Ligand Hydroxylation

First information on the mechanism of tyrosinase was obtained by studies referring to the hydroxylation of an aromatic, but non-phenolic part in the employed ligand framework, starting with the **XYL**-based system by Karlin *et al.* in the mid-eighties (Figure 1).[19] Following studies on different model systems by Tolman *et al.*, Réglier *et al.*, Schindler *et al.*, and again by Karlin *et al.* gave further insight into the mechanism of these ligand hydroxylation reactions, indicating that a $\mu\text{-}\eta^2\text{:}\eta^2$-peroxo/bis-$\mu$-oxo intermediate typically formed upon exposure of a Cu(I) precursor to dioxygen mediates electrophilic attack on the ligand framework.[11,19-27] These investigations provided significant information toward a mechanistic understanding of tyrosinase activity.

Given the fact that tyrosinase specifically mediates the monooxygenation of phenols (and not non-phenolic aromatics), a further development

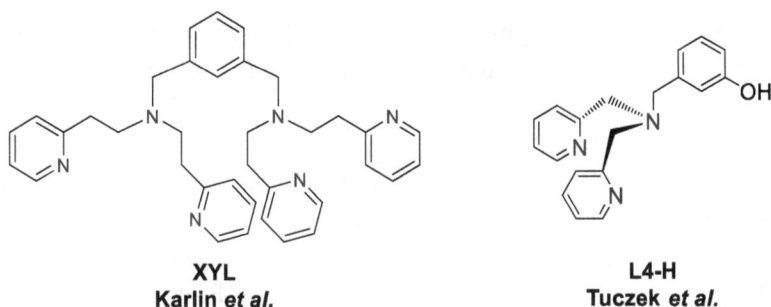

XYL
Karlin *et al.*

L4-H
Tuczek *et al.*

Figure 1.   The **XYL** system by Karlin *et al.*, which is able to perform the hydroxylation of the aromatic ring in the ligand backbone and the **L4-H** system that performs hydroxylation of the phenolic residue.[19,28]

of these systems toward mediating the hydroxylation of a phenolic residue appended to the ligand framework appeared of interest. Apart from Karlin *et al.* and Itoh *et al.*, only few studies have been published on this issue so far.[29–32] In 2015, our workgroup presented a model system exhibiting the described reactivity. The tridentate **L4-H** system is composed of a bis(2-pyridylmethyl)amine unit (Figure 1), which is complemented with a phenol residue to a tertiary amine.[28] In the presence of dioxygen at low temperatures, the mononuclear copper(I) complex of **L4-H** mediates an aromatic hydroxylation of the appended phenol, that is, the phenol residue in the ligand backbone is *ortho*-hydroxylated and oxidized to the corresponding CuL4quinone derivative upon reaction of the Cu(I) precursor with dioxygen at low temperatures.[28]

Apart from tyrosinase models that are able to hydroxylate an aromatic part of the ligand framework,[6,7,19–33] model systems have been developed, which are able to convert external monophenolic substrates to the corresponding *ortho*-quinones in a stoichiometric or a catalytic fashion.[6,7,9,34–53] In the remainder of this review we will focus on systems converting external substrates.

B. Model Systems Exhibiting Reactivity Toward External Substrates

Synthetic model systems with binuclear copper sites that catalyze the conversion of external monophenols to the corresponding *ortho*-quinones have been studied for a long time.[6,7,54] Bulkowski *et al.* took the first steps in the field of catalytic model systems for tyrosinase in 1985 with the development of the necessary conditions for a catalytic monooxygenation of phenolic substrates in an enzyme-like fashion.[35] He used copper(I) complexes supported by dinucleating, macrocyclic tetraamine ligands (Figure 2) as catalysts. For the conversion of phenols to *ortho*-benzoquinones in the presence of these complexes and dioxygen, an excess of the external monophenol (100 eq.) and triethylamine (200 eq.) was employed. The base was used for deprotonation of the phenol, leading to a better coordination to the copper ions, a step that is necessary for the following *ortho*-hydroxylation.[35]

In the following, the conditions established by Bulkowski *et al.* were applied in further catalytic studies and were improved by Réglier and coworkers in 1990 with their Cu(I)$_2$(**BiPh(impy)$_2$**) system (Figure 2).[36]

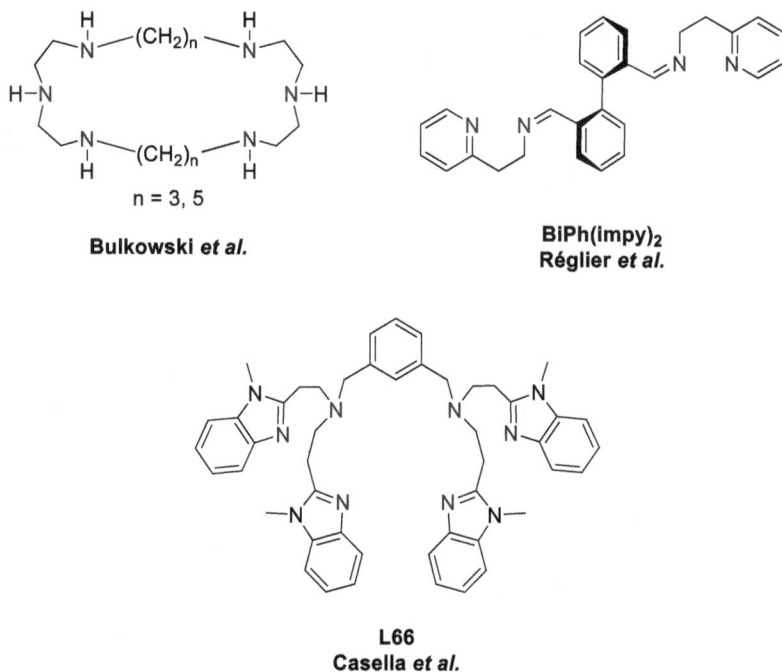

n = 3, 5

Bulkowski et al.

BiPh(impy)$_2$
Réglier et al.

L66
Casella et al.

Figure 2. Early model systems with the ability to catalytically convert monophenols to *ortho*-quinones.[35,36,55]

The reaction of this copper(I) catalyst with 2,4-di-*tert*-butylphenol (2,4-DTBP-H, 100 eq.) and 200 equivalents of triethylamine (from here on called *Bulkowski–Réglier conditions*) in the presence of dioxygen led to the formation of the corresponding 3,5-di-*tert*-butyl-*ortho*-quinone (3,5-DTBQ) with a turnover number of 16 after one hour.[36]

In 1991, Casella and coworkers established a further system supported by the ligand **L66**, which was able to mediate the stoichiometric as well as catalytic tyrosinase-like reactivity of external phenolic substrates.[18,55] Importantly, in the year 2000, the μ-η^2:η^2-peroxo intermediate of the Cu(I)$_2$**L66** system could be generated reversibly and shown to be the hydroxylating agent in monooxygenation reactions (Figure 2).[56,57] Following studies by Garcia-Bosch, Karlin, and Solomon as well as Herres-Pawlis *et al.* gave further important insight into the mechanism of the tyrosinase reaction.[38,43,58–61]

In most of the published studies on catalytic monooxygenations of phenolic substrates, the combination of a significant excess of the substrate and base was applied, and the obtained results indicate that the excess of the added base is indeed important for the catalytic activity.[6,7,35-43,55,56,58-61] By the addition of neutral phenols as substrates, phenoxyl radicals are generated and only unphysiological radical coupling products were obtained.[6] This constitutes a remarkable difference to the enzymatic system, which mediates the monooxygenation of phenols in the absence of an external base and without side-products like C–C-coupled molecules.

In order to increase the catalytic activity regarding the formation of *ortho*-quinones and realize more biomimetic model systems, a range of model systems was established by several workgroups during the last years. Depending on the employed ligand framework, characteristic differences in the monooxygenation activity toward various monophenols were observed. Scheme 2 depicts the possible aspects according to which these systems can be classified. An important distinction exists between structural model systems, which are primarily designed to stabilize copper–oxygen intermediates, and functional models with the ability to stoichiometrically or catalytically mediate the tyrosinase reaction.[11] Regarding

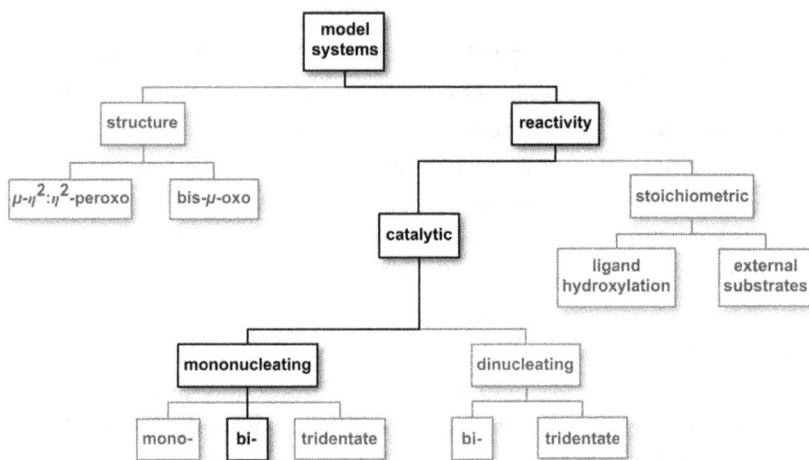

Scheme 2. Different aspects of tyrosinase model chemistry and scope of this review (black).

the latter category, copper complexes supported by mono- or dinucleating ligands can be distinguished in turn. Moreover, simple copper salts have been found to catalyze the monooxygenation of phenols.[48,49,62] Describing all of these systems would be beyond the scope of this chapter, so only findings regarding catalytic systems supported by mononucleating, bidentate N-donor ligands are discussed in the following.

III. Catalytic Tyrosinase Models with Bidentate Ligands

A. Mechanistic Cycle of the Tyrosinase-like Monooxygenation of External Substrates

In 2005, Stack et al. reported the first Cu(I) complex supported by a bidentate ligand exhibiting tyrosinase activity toward external substrates in a stoichiometric fashion (Figure 3).[34,47] By exposure to dioxygen, the Cu(I) **DBED** complex (**DBED** = N,N'-di-tert-butyl-ethylenediamine) was shown to form a μ-η^2:η^2-peroxo dicopper(II) species. Addition of

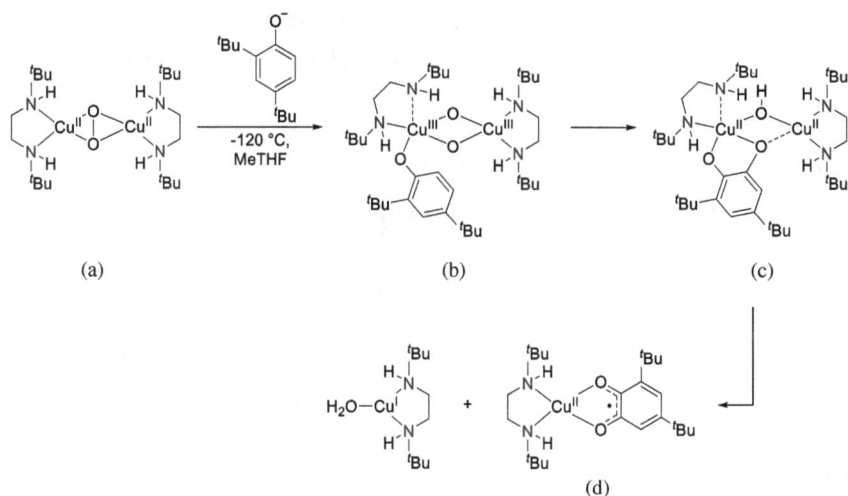

Figure 3.   The first mononucleating model system with tyrosinase activity, developed by Stack et al. and (a) the μ-η^2:η^2-peroxo dicopper(II) complex, which is formed upon coordination to copper(I) and oxygenation at low temperatures, (b) the ternary adduct between a bis-μ-oxo dicopper(III) complex with bound phenolate, (c) the catecholato adduct, and (d) the resulting copper(II)-semiquinone complex.[34,47]

phenolate led to isomerization to a bis-μ-oxo dicopper species, in which attack on the bound substrate occurs, leading to a bound catecholato species (and subsequently to a copper(II)-semiquinone complex). Importantly, this system provided the first evidence for a "ternary intermediate" (Cu + O_2 + substrate), which gave additional hints on how this reaction might occur in the enzyme. In 2014 Lumb *et al.* showed that the copper(I) **DBED** system also exhibits catalytic tyrosinase activity if a slightly over-stoichiometric amount of ligand is used. In this case the diamine ligand also adopts the role of a base.[48] Applying this protocol, Lumb *et al.* described the catalytic conversion of several monophenols to the corresponding *ortho*-quinones and other functionalized organic products, respectively.[50–53]

The first catalytic tyrosinase model complex supported by a bidentate ligand was presented in 2010.[37] The employed ligand $\mathbf{L_{py}1}$ (Figure 4) represents one-half of the dinucleating **BiPh(impy)₂** ligand (see above and Figure 2) and contains a combination of a *tert*-butyl-terminated imine and a pyridine.[36,37] Employing Bulkowski-Réglier conditions, the Cu(I) complex in the presence of dioxygen mediates the formation of 3,5-DTBQ with a TON of 22 after six hours.[37] This discovery was the starting point for the development of other mononuclear catalytic model systems of tyrosinase supported by bidentate *N*-donor ligands. The focus of this research was mainly put on increasing the activity and the stability of the corresponding catalysts.[37,39,40,42,44,45,63]

Based on comprehensive spectroscopic studies, a catalytic cycle was derived for the $\mathbf{L_{py}1}$ system (Scheme 3).[37] By addition of two equivalents of 2,4-DTBP-H and triethylamine to the Cu(I)$\mathbf{L_{py}1}$ complex (**1**), Cu(I) complex **2** is formed, which could be isolated and analyzed using UV/vis and NMR spectroscopy. Through oxygenation of **2** a band at 340 nm appears in the UV/vis spectrum indicating the formation of a μ-catecholato-μ-hydroxo intermediate (**4**). Furthermore, an HCl quench following the oxidation of **2** leads to 3,5-DTBQ, as well as 3,5-di-*tert*-butylcatechol (3,5-DTBC-H₂), deriving from the catecholato adduct in **4**. Under Bulkowski-Réglier conditions, **4** releases the bound catecholato adduct as 3,5-DTBQ. The protonation of the hydroxide group of **4** through protonated triethylamine initiates the formation of 3,5-DTBQ, indicating that triethylamine acts as a proton shuttle. The release of 3,5-DTBQ and water

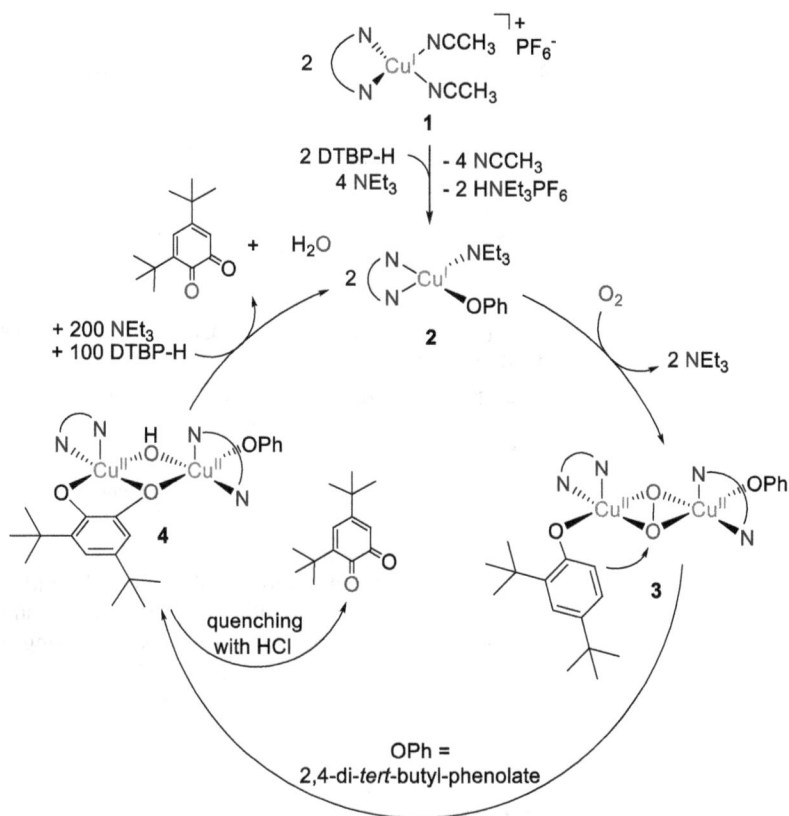

Scheme 3. Catalytic cycle for the monophenolase reaction mediated by bidentate N-donor model systems.[7,37,39,40,42,45]

results in a reduction of Cu^{II} to Cu^{I} and the reformation of **2**, thus closing the catalytic cycle.[37]

The isolation and characterization of the catecholato adduct **4** imply the presence of a ternary phenolato-peroxo complex **3**, which, however, could not be detected in the $L_{py}1$ system. Nevertheless, formation of the $\mu\text{-}\eta^2\text{:}\eta^2$-peroxo species could be evidenced at 185 K in acetone using UV–vis spectroscopy, showing that this species is also mechanistically relevant.

In 2014 Lumb *et al.* used $L_{py}1$ derivatives for the conversion of 4-*tert*-butylphenol to the corresponding coupled quinone, which was mediated

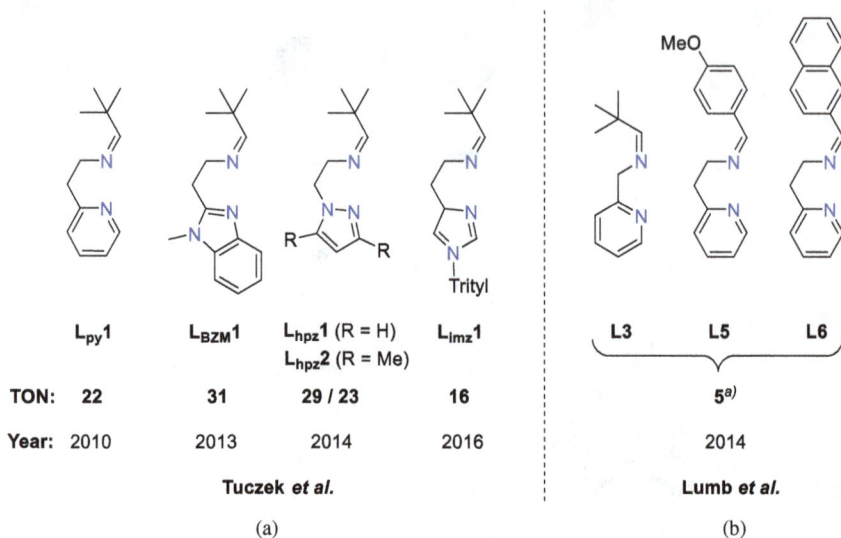

Figure 4. Model systems with an imine moiety and an *N*-heterocycle and the obtained TONs under Bulkowski–Réglier conditions with 2,4-DTBP-H as substrate and a catalyst concentration of 500 μmol/L. (a) The TONs were obtained using 0.1 mmol catalyst, 10 eq. of 4-tert-butylphenol and 5 eq. of NEt$_3$.[37,39,40,44,48]

by $L_{py}1$ derivatives.[48] The substitution of the *tert*-butyl moiety of $L_{py}1$ with 4-methoxybenzene and naphthalene led to **L5** and **L6**, respectively. For ligand **L3** the alkyl chain of $L_{py}1$ was shortened to only one CH$_2$-unit (Figure 4). In this study a 1:10 and thus higher catalyst-to-substrate ratio was applied and the oxygenation was performed in the absence and presence of 4 Å molecular sieves. These conditions led to product yields of 57–58% (TON ≈ 3) without desiccant, and 93–96% (TON ≈ 5) with the use of molecular sieves, respectively.[48]

B. Substitution of Pyridine in the $L_{py}1$ Ligand by *N*-Heterocycles

To learn more about electronic and structural influences on the reactivity of the $L_{py}1$ system toward phenolic substrates, and eventually increase its catalytic activity, the pyridine moiety was first substituted by other heterocyclic groups. Notably, previous investigations had shown that Cu(I) complexes supported by ligands, composed of two pyridine (**DPM**) and two *tert*-butyl-terminated imine functions (**L2**), are catalytically inactive.[37,64] Therefore, the combination of an imine unit and a heterocycle appeared to be superior to symmetric ligands composed of two heterocycles/two imine units (see below).

In pursuit of the stated goals, analogs of $\mathbf{L_{py}1}$ were established, which in particular contain (i) a benzimidazole ($\mathbf{L_{bzm}1}$), (ii) a pyrazole ($\mathbf{L_{hpz}1}$ and $\mathbf{L_{hpz}2}$), or (iii) an imidazole ($\mathbf{L_{imz}1}$) moiety instead of the pyridine ring (Figure 4).[37,39,40,44] All of these systems showed catalytic monophenolase activity toward 2,4-DTBP-H.[39,40,44] By substitution of the pyridine moiety of $\mathbf{L_{py}1}$ with pyrazole ($\mathbf{L_{hpz}1}$) and 3,5-dimethylpyrazole ($\mathbf{L_{hpz}2}$), respectively, the catalytic activity of the systems increased from a TON of 22 to 29 and 23, respectively.[40] The substitution of pyridine with benzimidazole ($\mathbf{L_{bzm}1}$) even resulted in a TON of 31, whereas the system with an imidazole as *N*-heterocycle only exhibited a TON of 16.[39,44] Thus, using pyrazole and benzimidazole instead of pyridine increases the catalytic activity whereas imidazole lowers the TON. In contrast to the parent $\mathbf{L_{py}1}$ system it was not possible to detect a $\mu\text{-}\eta^2\text{:}\eta^2$-peroxo intermediate for any of these modified systems.[37,39,40,44]

The obtained experimental results can be rationalized by considering the bonding properties of the employed *N*-heterocyclic groups. In order to mediate an electrophilic substitution (S_E) on the arene ring of the coordinated substrate, the side-on-bound peroxo ligand should transfer as much electron density as possible into the two copper centers. This will in turn be facilitated by employing weakly σ-donating and/or strongly π-accepting co-ligands. As a measure of σ-donor strength, the pK_a value of the bound *N*-heterocycle (in its protonated form, cf. Table 1) may be employed.[65] Inspection of these values explains that replacement of pyridine by, for example, pyrazole increases the catalytic efficiency. A further increase of activity should be possible by going to triazole (see below). On

Table 1. pK_a values of imines and different *N*-heterocycles (in their protonated form).[65,66,67]

	imine	imidazole	benzimidazole	pyridine	pyrazole	triazole
pK_a	7–8	7	5.6	5.2	2.5	1.17

the other hand, exchange of pyridine by imidazole should lower the catalytic activity as σ-donation is increased, again in agreement with the experimental observation.

Of particular interest is the large activity increase observed upon replacing pyridine by benzimidazole. Based on the pKa values, this substitution should have little effect. However, benzannulation lowers both the HOMO and the LUMO of imidazole.[67,68] While the former explains the decrease of σ-donation (and concomitant decrease of pK_a) upon going from imidazole to benzimidazole (cf. Table 1), the latter acts to increase the π-backbonding capability, that is, benzimidazole should be more backbonding than imidazole. Probably, the combination of both effects makes benzimidazole superior to pyridine when employed as a donor group in copper monooxygenation catalysts. This has already been exploited by Casella whose **L66**- and **L6**-supported tyrosinase models, for example, contained benzimidazoles as terminal *N*-donors (and not pyridines; see above).[55,56]

C. Substitution of Imine in the $L_{py}1$ Ligand by *N*-heterocycles: Symmetric Bidentate Ligands

After replacement of the pyridine heterocycle in the ligand backbone by other *N*-heterocycles, it was of interest to check whether the imine moiety of the $L_{py}1$ ligand can be replaced by an *N*-heterocycle as well. First, symmetric ligands were considered. As the **DPM** system had been found to be catalytically inactive (see above),[37] the focus was first put on (substituted) pyrazoles as *N*-heterocycles.[42] Specifically, Cu(I) complexes were prepared using the ligands 1,1'-methylenebis-1 *H*-pyrazol (**BPM**), 1,1'-methylenebis(3-methyl-1*H*-pyrazole) (**mBPM**) and 1,1'-methylenebis(3,5-di-methyl-1*H*-pyrazole) (**dmBPM**; Figure 5).[42] By investigating the model systems regarding the monooxygenation of 2,4-DTBP-H, catalytic activity could be detected for all three model systems. Specifically, the **BPM-**, **mBPM-**, and **dmBPM-**supported systems mediated the conversion to 3,5-DTBQ with TONs of 21, 19 and 11, respectively. Besides 2,4-DTBP-H various other substrates were tested, for example small substrates like phenol (P-H), substrates containing electron-drawing residues such as 4-methoxyphenol

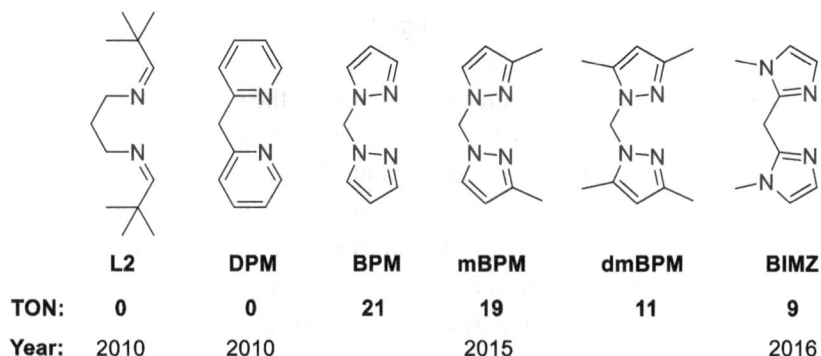

	L2	**DPM**	**BPM**	**mBPM**	**dmBPM**	**BIMZ**
TON:	0	0	21	19	11	9
Year:	2010	2010		2015		2016

Figure 5. Symmetric model systems for tyrosinase and their catalytic activity under Bulkowski-Réglier conditions *vs.* 2,4-DTBP-H and a catalyst concentration of 500 μmol/L.[37,42,44,64]

(4-MeOP-H), or the biomimetic substrate *N*-acetyl-L-tyrosine ethyl ester monohydrate (NATEE). The monophenolase reaction was catalyzed by the model systems for all substrates, albeit with different TONs.[42] These findings allowed to derive substrate-specific factors influencing their catalytic monooxygenation (see below). Interestingly, it was possible to detect the μ-η^2:η^2-peroxo intermediate for the **dmBPM** system at –90°C in acetone, whereas this was not possible for the **BPM** and **mBPM** systems.[42] This strongly indicates a steric influence on the stability of the peroxo intermediate as well as on the reactivity of the model systems (this point is discussed in more detail below).

By substitution of the pyrazole moieties with *N*-methylimidazoles, the **BIMZ** ligand was obtained (Figure 5). Although a monophenolase activity towards 2,4-DTBP-H was detected, only a small TON of 9 was observed.[44] Note that the increase of catalytic activity upon going from **DPM** to the pyrazole-based ligands **BPM**, **mBPM** and **dmBPM** corresponds to the results obtained with the asymmetric $\mathbf{L_{het}X}$ ligands (het = *N*-heterocycle, X = 1 or 2; cf. previous section), where replacement of $\mathbf{L_{py}1}$ by the (less electron-donating) $\mathbf{L_{hpz}1}$/$\mathbf{L_{hpz}2}$ ligands was found to increase the catalytic activity of the corresponding Cu(I) complexes. Moreover, replacement of the pyrazole moieties in **BPM** by (more electron-donating) imidazoles led to a decreased catalytic activity of the Cu(I) complex supported by the **BIMZ** ligand, again in line with the results obtained for the asymmetric ligands.

Overall, however, the catalytic activities of the Cu(I) complexes supported by the symmetric ligands are conspicuously lower than those of their counterparts with asymmetric ligands. This result was surprising as it is not compatible with the bonding properties discussed so far: based on the corresponding pK_a values, all of the employed heterocycles are weaker σ-donors than imine (cf. Table 1). From a purely electronic point of view, replacement of imine by these moieties thus should *increase* (and not *decrease*) the catalytic activity. The complete catalytic inactivity of the **DPM** complex was even more striking; based on the activity sequence of the asymmetric systems it should lead to a TON between that of the imidazole and the pyrazole system.

In an attempt to understand the described findings, the formation and possible constitution of the ternary intermediate **3** (cf. Scheme 3) are considered in more detail. Note that in a stoichiometric reaction mode (upon which the mechanism depicted in Scheme 3 is based) the substrate is bound to the Cu(I) precursor *before* oxygenation. In a catalytic run, however, the substrate is added to the Cu(I) precursor $[Cu(\mathbf{L})(CH_3CN)_2]^+$ ($\mathbf{L}$ = bidentate ligand) and will bind to the $\mu\text{-}\eta^2\text{:}\eta^2$-peroxo complex formed upon oxygenation of the latter, generating ternary intermediate **3**. In order to get hydroxylated, the substrate has to bind within the Cu_2O_2 plane (Scheme 4).[6]

Such a configuration, in which the peroxide σ^* orbital overlaps with the π-system of the arene ring[6] can in principle be reached by two pathways which correspond to an associative and a dissociative mechanism. In the *associative* scheme initial binding of the substrate proceeds axially to one of the Cu(II) centers of the planar $\mu\text{-}\eta^2\text{:}\eta^2$-dicopper complex; *e.g.* Cu^B

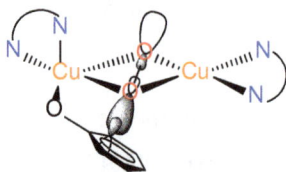

Scheme 4.   Overlap of a phenolate π-type orbital and the σ^* orbital of the peroxo ligand.[6]

associative pathway:

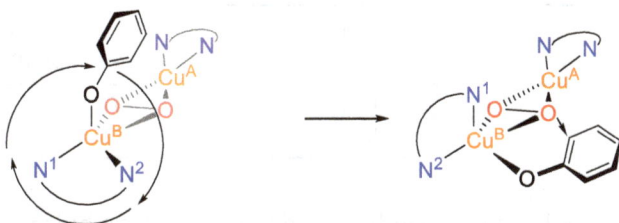

dissociative pathway:

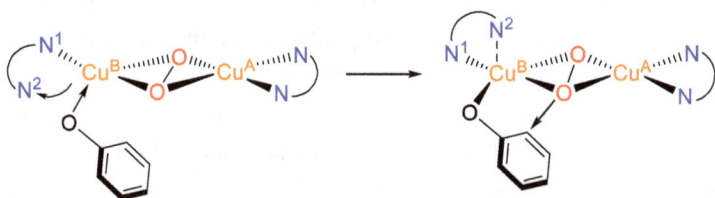

Scheme 5.   Associative and dissociative pathways of the substrate coordination for a monooxygenation reaction mediated by a catalytic copper complex supported by (top) symmetric and (bottom) asymmetric ligands.

(cf. Scheme 5 left). To reach an equatorial coordination, the ligand sphere around Cu^B has to rotate around an axis connecting the two copper centers (cf. Scheme 5, top). Such an axial $\rightarrow$ equatorial rearrangement has also been discussed for the enzyme.[4,6,69]

In the frame of a *dissociative* mechanism, equatorial coordination of the substrate in the Cu_2O_2 plane occurs after dissociation of one of the two donor groups of the bidentate ligand from the peroxo complex (Scheme 5. bottom). Rebinding of that donor after coordination of the substrate at the free equatorial position, may subsequently occur in axial position or not depending on its donor/acceptor strength (if not, the bidentate ligand would act as a hemilabile ligand).[70] In the context of this (dissociative) pathway, the function of a comparatively strong donor (such as imine) would be to keep the bidentate ligand bound to the peroxo dicopper complex and the ternary intermediate, respectively. On the other hand, such a dissociative scenario would be mechanistically unfavorable for a symmetric bidentate ligand with *two* comparatively weak donors as the ternary

Table 2. Observed TOFs (min^{-1}) after 15 min for different model systems upon catalytic conversion of 2,4-DTBP-H.[37,39,40,42,44]

	$L_{py}1$	$L_{bzm}1$	$L_{hpz}1$ ($L_{hpz}2$)	$L_{imz}1$	BPM (mBPM/dmBPM)	BIMZ
TOF (0–15 min)	0.41	0.84	0.85 (0.62)	0.38	0.27 (0.24/ 0.22)	0.17

intermediate would be less stabilized and total loss of the bidentate ligand may occur. It is thus conceivable that mixed ligands with one imine group (as a comparatively strong donor) and one (weaker coordinating) *N*-heterocycle follow a dissociative pathway for the binding of the substrate whereas symmetric ligands with two comparatively weak donors follow an associative pathway. The latter pathway is *a priori* kinetically disfavored compared to the former due to the fact that equatorial coordination of the substrate only occurs after an axial → equatorial rearrangement whereas this is not necessary within the dissociative scheme. This mechanistic difference may explain the overall lower activity of the symmetric ligands, which is also evident from lower observed TOFs as compared to their asymmetric, imine-containing analogs (cf. Table 2). The total lack of catalytic activity observed for the **DPM** system, finally, may be associated with the fact that this is the only symmetric ligand that is composed of two six-membered rings as opposed to the other symmetric ligands that contain two *N*-heterocycles with five-membered rings. This renders the bite angle of DPM smaller than that of the latter systems, which in turn could become relevant, *e.g.*, in the course of the axial → equatorial rearrangement.

D. Hybrid Ligands Composed of Different *N*-Heterocycles

In view of the fact that the imine function of $L_{py}1$ and related systems is unstable against hydrolysis, which might be one reason for limited TONs of the corresponding copper catalysts,[37,44] it also appeared of interest to replace it by hydrolytically more stable oxazoline units. Combination with imidazole groups led to the three asymmetric ligands $L_{OL}1$, $L_{OL}2$ and $L_{OL}3$ (Figure 6).[44] Three different substituents; i.e., phenyl, methyl and *tert*-butyl, provided electronic and steric variation of the ligands.

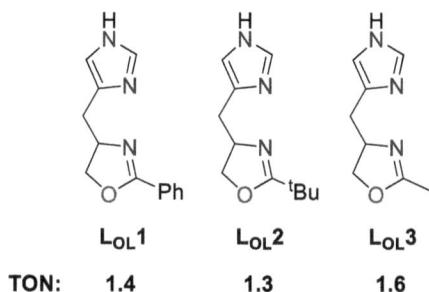

$L_{OL}1$ $L_{OL}2$ $L_{OL}3$

TON: 1.4   1.3   1.6

Figure 6. The $L_{OL}X$ model systems with the ability to mediate the tyrosinase reaction in a stoichiometric fashion under Bulkowski-Réglier conditions *vs.* 2,4-DTBP-H and a catalyst concentration of 500 μmol/L.[44]

Investigation of the catalytic activity, however, showed that the derived Cu(I) complexes are inactive regarding the catalytic monooxygenation of phenols and only mediate the conversion of 2,4-DTBP-H in a stoichiometric fashion.[44] This aspect could be explained with the influence of structural factors like the bulky substituents of the oxazoline residues and a lower flexibility of the ligand backbone. The oxazoline units, especially in case of the $L_{OL}1$ ligand, are more sterically demanding than the imine function.[44]

In order to further explore whether bidentate ligands for copper monooxygenase models can be assembled in a modular fashion, a hybrid of the (catalytically inactive) dipyridylmethane (**DPM**) and the (catalytically active) bispyrazolylmethane (**BPM**) system was investigated in 2018. Specifically, the ligands pyrazolmethylpyridine (**PMP**) and 3,5-dimethylpyrazolmethylpyridine (**dmPMP**) were prepared (Figure 7),[37,42,45] and corresponding copper(I) complexes were investigated regarding their ability to catalyze the oxygenation of several monophenolic substrates. Regarding 2,4-DTPB-H these complexes exhibited TONs of 14 for **PMP** and 11 for **dmPMP**. Remarkably, the TON of the **PMP** system is intermediate between that of the catalytically inactive **DPM** (TON = 0) and the medium-to-highly active **BPM** model system (TON = 21),[45] confirming the foregoing hypothesis. Again, the methyl-substituted **dmPMP** system exhibits a lower activity towards external substrates than its parent, unsubstituted analog. The

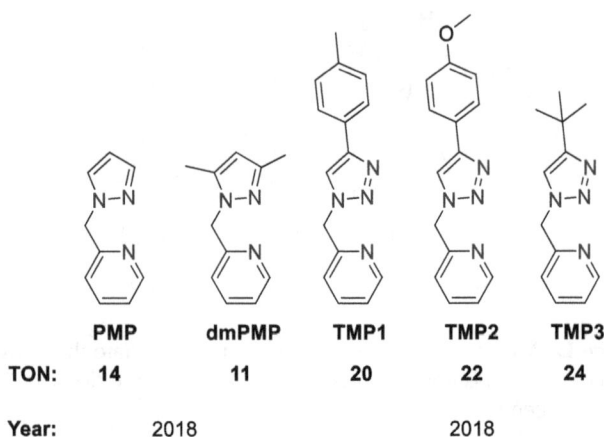

	PMP	dmPMP	TMP1	TMP2	TMP3
TON:	14	11	20	22	24

Year:	2018	2018

Figure 7. Ligands containing different *N*-heterocycles and the TON of the corresponding model systems under Bulkowski-Réglier conditions with 2,4-DTBP-H as substrate and a catalyst concentration of 500 μmol/L.[45,63]

μ-η^2:η^2-peroxo intermediate could be observed for both complexes upon exposure to dioxygen at low temperatures, although with low yields (2.5% for the **PMP** and 3.3% for the **dmPMP** system).[45]

Interestingly, crystals grown from a solution of [Cu(**PMP**)(MeCN)$_2$] PF$_6$ provided evidence for the formation of the homoleptic complex [Cu(**PMP**)$_2$]PF$_6$. For the **dmPMP** system the analogous complex was not isolated under these conditions, but it was possible to obtain it by direct synthesis.[45] These findings suggest the existence of an equilibrium between the heteroleptic and the homoleptic complex in solution, potentially rendering the mechanism of the catalytic monooxygenation reaction more complicated. Density functional theory (DFT) was used to gain more insights into the behavior of the two systems in solution. Importantly, these calculations supported the presence of an equilibrium between the heteroleptic and the homoleptic complex in solution for both systems.[45] Furthermore, investigation of the **PMP** systems led to single crystals of a dinuclear complex in which two Cu(II)-**dmPMP** units are bridged by a fluoro ligand deriving from the counter ion PF$_6$.[45] This might correspond to one of the decay products of **PMP**-type catalysts after prolonged reaction times and give a hint on possible deactivation pathways for this type of catalysts (see below).

As the electron-poor N-donor pyrazole in ligands such as **PMP** and **BPM** led to medium-to-high catalytic activities (see above), using even less electron-donating N-heterocycles such as triazoles was supposed to further increase the reactivity towards external substrates (cf. Table 1).[42,45,63] Exchanging the pyrazole group in **PMP** by triazole led to the **TMP** family of ligands containing a tolyl (**TMP1**), anisole (**TMP2**) and *tert*-butyl (**TMP3**) residue at the triazole's C4 position (Figure 7). The ability of these systems to mediate the monophenolase reaction was investigated using 2,4-DTBP-H, 3-TBP-H (3-*tert*-butylphenol) and 4-MeOP-H (4-methoxyphenol) as substrates. In fact, the corresponding model systems exhibited high catalytic activities towards 2,4-DTBP-H with TONs of 20, 22 and 24 for **TMP1**, **TMP2** and **TMP3**, corroborating the inverse relationship between the σ-donor strength of the supporting ligand and the catalytic monooxygenation activity of the derived Cu(I) complex (see above). Employing the **TMP3** model system, the highest TON (48) yet observed for the copper-mediated monooxygenation of a phenolic substrate (4-MeOP-H) was achieved.[63] Interestingly and in contrast to the previous model systems, the influence of the residue on the triazole heterocycle was found to be small. Thus, the **TMP3** system with the most sterically demanding *tert*-butyl residue exhibited the highest catalytic activity. The difference towards the methyl-substituted systems of **BPM**, **PMP**, and $\mathbf{L_{hpz}}$ is that the C3 and C5 positions are unsubstituted for the **TMP** systems and the residue is bound in C4 position.[42,45] This results in a larger distance to the peroxo ligand, which leads to a smaller steric influence. Therefore, although the residues for the **TMP** systems are larger than for the previous systems, their steric influence is small. As for the **PMP** model systems a $\mu\text{-}\eta^2\text{:}\eta^2$-peroxo intermediate was also detected for the **TMP** model systems, although only in small yields of 1.1% to 2.6%.[63]

Concluding this chapter, it should be mentioned that recently Herres-Pawlis *et al.* presented another model system with a bidentate N-donor ligand. Supported by a hybrid guanidine ligand the corresponding copper complex forms a bis-μ-oxo dicopper(III) species when exposed to oxygen at low temperatures. The copper-oxygen species catalytically converts different phenolic substrates into *ortho*-quinones with high product yields.[61]

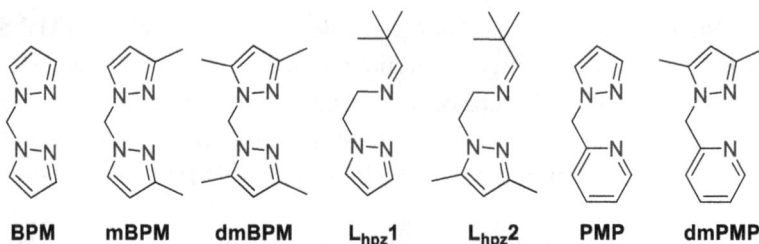

Figure 8. Different bidentate ligands with and without methyl residues.[40,42,45]

Table 3. Catalytic activity of different **unsubstituted** model systems and their methyl substituted counterparts.[40,42,45]

Model system	TON *vs.* 2,4-DTBP-H	Decrease of activity *vs.* unsubstituted system (in %)
L$_{hpz}$1 (L$_{hpz}$2)	**29** (23)	21
BPM (mBPM/dmBPM)	**21** (19/11)	10/48
PMP (dmPMP)	**14** (11)	21

E. Steric Influence of Substituents on the Supporting Ligands

In the preceding sections the electronic influence on the catalytic activity of copper catalysts supported by bidentate ligands has been considered in detail. As already mentioned, the use of different residues altering the steric bulk of the ligand adds another parameter influencing the catalyst's activity. This first became evident when investigating the **L$_{hpz}$1** system, which exhibited a TON of 29 towards 2,4-DTBP-H (Figure 8): using the 3,5-dimethyl substituted derivate **L$_{hpz}$2**, decreased the activity by 21% (TON = 23, Table 3).[40] For the symmetric **BPM** systems analogous results were obtained throughout all examined substrates, with exception of phenol (P-H) and 4-MeOP-H; i.e., the unsubstituted **BPM** shows the highest activity towards external substrates, while the TON is lowered for the 3-methyl substituted **mBPM** and significantly decreased for the 3,5-dimethyl substituted **dmBPM**. The conversion of 2,4-DTBP-H to 3,5-DTBQ, *e.g.*, is mediated with a TON of 21 for **BPM**, while reduced by 10%/48% when using **mBPM/ dmBPM**, respectively. The same observation was made for the **PMP** systems: while the parent, unsubstituted system mediates the conversion to 3,5-DTBQ with a TON of 14, the activity is lowered by 21% for the **dmPMP** system, where a TON of 11 was obtained (Table 3).[45]

These findings can be explained by the fact that the presence of bulky residues in the ligand backbone disfavors the coordination of the substrate to the copper center in the peroxo intermediate (Scheme 3) as well as electrophilic attack of the $\mu\text{-}\eta^2\text{:}\eta^2$-peroxo ligand on the coordinated phenol in *ortho*-position (Scheme 3).[37] Thus, conversion of the substrate to the *ortho*-quinone is hampered and the system's catalytic activity reduced. On the other hand, steric shielding of the peroxo ligand stabilizes the copper-peroxo complex. Thus, detection of the $\mu\text{-}\eta^2\text{:}\eta^2$-peroxo intermediate was successful for the **dmBPM** system, while it could not be achieved for **BPM** or **mBPM**.[42] In agreement with this, the yield of the $\mu\text{-}\eta^2\text{:}\eta^2$-peroxo-intermediate detected for the **dmPMP** was larger than for the **PMP** system.[45] An important consequence of the lower reaction rate associated with steric shielding may be that the significance of side reactions, deactivating the catalyst, increases. One of these pathways is the reaction of the Cu(II) catecholato complex (Scheme 3) with water, which is liberated in the course of the catalytic cycle, leading to a bis-μ-hydroxo species.[48] Other possible decay products are generated by reactions with anions deriving from the counterion (cf. the μ-fluoro-bridged dinuclear Cu(II) **dmPMP**; see above).[45]

F. Variation of Substrates: Electronic and Steric Factors

Apart from the electronic and steric properties of the ligands, which have been considered in the previous sections, the product yields and corresponding TONs of copper-mediated monooxygenation reactions are significantly influenced by the nature of the substrate and its interaction with the catalyst. The substrates which we have employed to investigate the tyrosinase activity of our model systems range from small or low-activated (P-H, 4-MeP-H) to sterically demanding (2,4-DTBP-H, NATEE) and highly activated (4-MeOP-H) phenols.[7,37,42,44,45,63] In the following the different properties of the used substrates and their influence on the outcome of the monooxygenation reaction are briefly discussed.

P-H: R = H, R' = H   **Q:** R = H, R' = H
2,4-DTBP-H: R = 'Bu, R' = 'Bu   **3,5-DTBQ:** R = 'Bu, R' = 'Bu
4-MeP-H: R = H, R' = Me   **4-MeQ:** R = H, R' = Me
NATEE: R = H, R' = $C_7H_{12}NO_4$ **NADQEE:** R = H, R' = $C_7H_{12}NO_4$

4-MeOP-H: R = H, R' = OMe   **cQMeO:** R = OMe   **pQ**
3-TBP-H: R = 'Bu, R' = H   **cQTB:** R = 'Bu

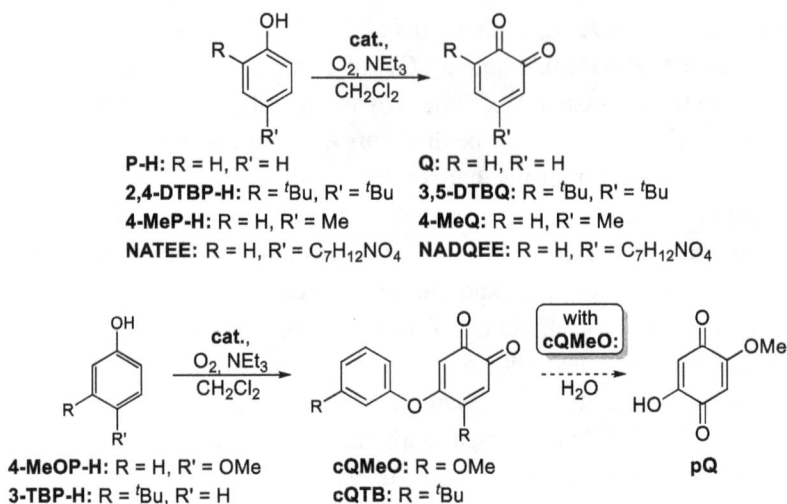

Scheme 6. The reactivity of the various investigated substrates.[37,42,48,49,63,72]

Although the substrate 2,4-DTBP-H carries two bulky *tert*-butyl residues and is therefore sterically hindered, it is commonly used in investigations of monophenolase activity.[7,44,45,63] Apart from increasing the steric bulk of the substrate the residues induce a +I-effect, thus increasing the electron-density in the aromatic system.[7,42,71] This favors the electrophilic attack by the peroxo ligand. Furthermore, the corresponding *ortho*-quinone (Scheme 6 top, λ_{abs} = 407 nm, ε = 1830 L mol^{-1} cm^{-1}) exhibits a high stability, hence isolation and characterization of this product is easy.[37,44,71] Nevertheless, when using 2,4-DTBP-H as a substrate the coupled biphenol 3,3',5,5'-tetra-*tert*-butyl-2,2'-biphenol (BP-H) is obtained as an unphysiological by-product. It is formed from phenoxyl radicals generated by H-atom transfer to the peroxo complex *via* an oxidative coupling reaction.[49,71]

An alternative to the "standard substrate" 2,4-DTBP-H is represented by 3-TBP-H, which reacts under catalytic conditions to the coupled quinone 4-(*tert*-butyl)-5-[3-(*tert*-butyl)phenoxy]cyclohexa-3,5-diene-1,2-dione (Scheme 6 bottom, λ_{abs} = 425 nm, ε = 898 L mol^{-1} cm^{-1}).[7,44,48] Here, only one *tert*-butyl residue is bound to the aromatic ring in *meta*-position, thus leading to a less sterically encumbered coordination to the copper

Table 4. Overview of the obtained TONs with different bidentate model systems[a] and different substrates.

	$L_{py}1$	$L_{hpz}1$ ($L_{hpz}2$)	BPM (mBPM/ dmBPM)	$L_{IMZ}1$	BIMZ	$L_{OL}1$–3	PMP (dmPMP)	TMP1 (TMP2/ TMP3)
2,4-DTBP-H	18	29 (23)	21 (19 / 11)	16	9	<1	14 (11)	20 (22 / 24)
3-TBP-H			22 (15 / 12)	20	6		25 (13)	20 (19 / 24)
4-MeOP-H			35 (10 / 15)	34	6		34 (33)	42 (45 / 48)
4-MeP-H			14 (12 / 10)					
P-H			9 (9 / 9)					
NATEE			18 (15 / 11)	25	0			

[a]**DPM** and **L2** are catalytically inactive and therefore not listed

center in comparison to 2,4-DTBP-H.[42,63] On the other hand, the electronic activation of the residue is smaller. Consequently for the most model systems with this substrate, similar TONs are observed as for 2,4-DTBP-H (Table 4).[42,44,63]

In contrast to the substrates considered so far, 4-MeOP-H does not contain bulky residues, but a methoxy group in *para*-position that activates the aromatic system due to the a +M-effect.[42,44,45,63] In accordance with the high activation and small steric bulk of the substrate, the observed TONs for this substrate are very high. Thus, TONs between 42 and 48 were observed for the **TMP** model systems (Table 4).[63] Under catalytic conditions 4-MeOP-H is converted into the coupled quinone 4-methoxy-5-(4-methoxyphenoxy)cyclohexa-3,5-diene-1,2-dione (Scheme 6 bottom, λ_{abs} = 418 nm, ε = 524 L mol^{-1} cm^{-1}).[44,48,73] In the presence of water the *para*-quinone 2-hydroxy-5-methoxy-1,4-benzoquinone is formed from the coupled quinone.[42,63,72]

When investigating the monophenolase activities of the symmetric **BPM** systems, three further substrates were tested.[42] The *para*-methyl substituted 4-MeP-H is slightly activated due to the +I-effect of the methyl group. The small activation leads to low *turnover frequencies* (TOFs) and TONs (14-10 for **BPM** systems, Table 4), despite the fact that the steric interaction between the substrate and the copper is small. In addition to the monophenolase reaction a side reaction occurred during the catalysis, reducing the substrate concentration and thus the possible product yield. Presumably, under catalytic conditions a copper semiquinone complex is formed and subsequently hydrolyzed in the presence of water, that itself is generated during the catalysis.[42] The formation of the 4-methylquinone is associated with an absorption band at 445 nm (Scheme 6 top, $\varepsilon = 1400$ L mol^{-1} cm^{-1}).[42] The smallest and least sterically hindered substrate is P-H; the corresponding *ortho*-benzoquinone exhibits an absorption band at 398 nm (Scheme 6 top, $\varepsilon = 1417$ L mol^{-1} cm^{-1}).[74] The lack of bulky residues, however, leads to two disadvantages. On the one hand, the phenol is not activated and therefore the resulting TOF is small. On the other hand, the unsubstituted aromatic C-H groups provide a large number of possible binding sites for side-reactions such as oxidative coupling or polymerization reactions.[49,51,75] Indeed for the **BPM** supported systems only TONs of 9 and low TOFs were obtained (Table 4).[42]

The substrate NATEE, finally, is closely related to the natural substrate L-tyrosine and therefore the best choice from a biomimetic point of view. Under catalytic conditions it is converted into *N*-acetyl-L-dopaquinone ethyl ester (NADQEE) with an absorption band at 392 nm (Scheme 6 top, $\varepsilon = 1417$ L mol^{-1} cm^{-1}).[76] With this substrate for the **BPM** model systems as well as for **L$_{imz}$1** medium to high TONs were observed using UV/vis spectroscopy, which could be attributed to the +I-effect of the residue and the fact that the residue is bound in *para*-position and therefore substrate coordination to the copper center is not sterically hindered (Table 4).[42,44] However, the absorption band corresponding to the *ortho*-quinone decreased after a few hours and the isolation *via* an HCl quench failed as well. Presumably the product is unstable in the presence of water, that originates from the catalytic cycle and the substrate itself.[42,44]

To conclude, it is possible to mimic tyrosinase activity with our model systems employing a large range of monophenols. It was found that especially the activation of the aromatic system through the electronic

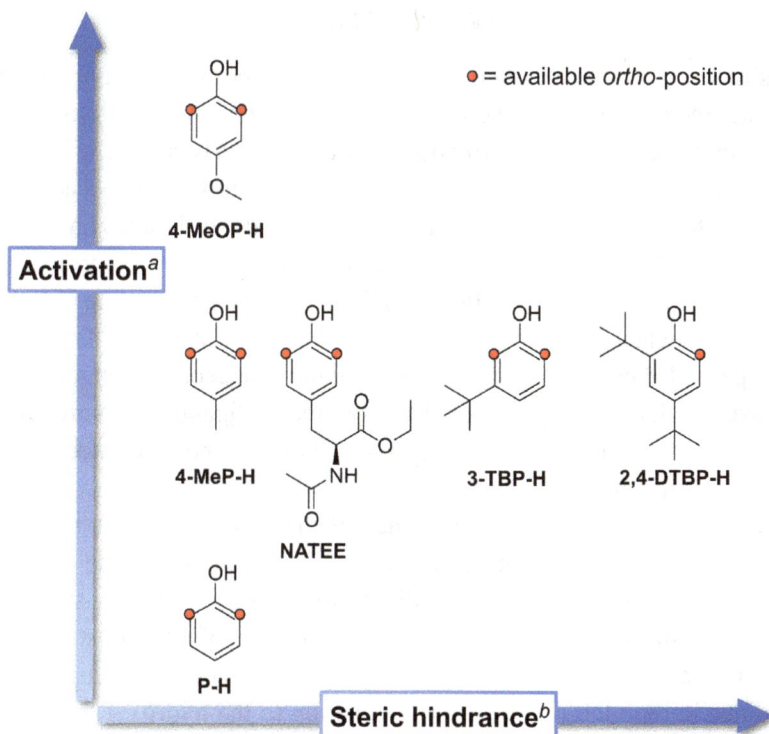

Scheme 7.   a: electronic activation of the aromatic system by the +I / +M-effect of the residues, b: steric hindrance of the residues upon coordination on the copper center.

influence of the residues greatly influences the resulting turnover number.[7,42,44,45,63] Furthermore, due to steric hindrance it is favorable to use substrates without residues in or in vicinity to the *ortho*-position to the hydroxyl group. These findings are graphically summarized in Scheme 7, showing the interplay between electronic activation and steric bulk. As 4-MeOP-H is not sterically encumbered and exhibits a high activation of the aromatic system induced by the +M-effect of the methoxy group, usually the highest TONs are observed for this substrate (Table 4).[42,44,45,63] For the biomimetic substrate NATEE, on the other hand, a high conversion to the corresponding *ortho*-quinone is also observed; however, this system suffers from the decay of NADQEE in the presence of water.[42,44]

IV. Summary

Tyrosinase is a dinuclear copper protein that converts L-tyrosine to L-dopaquinone. Besides structural and mechanistic investigations of the natural enzyme relevant information about the underlying chemistry can be obtained using small-molecule model systems. In the last decade several catalytic tyrosinase models with bidentate *N*-donor ligands have been established. Through detailed investigations of these systems, correlations between the properties of the ligand and the substrate on the one side and the observed catalytic activity on the other could be obtained. An important aspect of these correlations is the σ-donor/π-acceptor strength of the employed *N*-donor groups. Moreover, asymmetric ligands are more effective than their symmetric counterparts, which can be attributed to the kinetics of substrate binding.

Apart from the electronic influence of the *N*-donor groups the presence of bulky residues in the ligand backbone also influences the catalytic activity of the system. While stabilizing the peroxo complex, sterically demanding substituents may inhibit coordination of the substrate and electrophilic attack by the peroxo group. On the other hand, small steric hindrance in the vicinity of the *ortho*-position to the hydroxyl group and an increased electron density within the aromatic system (*e.g.*, through a +M-effect of a methoxy group in 4-position) facilitate bonding of a phenolic substrate to the copper-peroxo intermediate and subsequent electrophilic attack by the peroxo group. The reactivity of the presented tyrosinase model systems towards external monophenols thus is well understood. Remaining challenges to be adressed refer to increasing the turnover numbers, which are still low, and reducing the amount of base needed for a catalytic activity of the investigated model systems.

V. Acknowledgment

The authors would like to thank M. Rolff, J. Schottenheim, J. Hamann, F. Wendt, B. Herzigkeit, R. Schneider and other present and former members of the copper workgroup for their contributions to the research described in this review.

VI. References

1. Solomon, E. I.; Heppner, D. E.; Johnston, E. M.; Ginsbach, J. W.; Cirera, J.; Qayyum, M.; Kieber-Emmons, M. T.; Kjaergaard, C. H.; Hadt, R. G.; Tian, L. *Chem. Rev.* **2014**, *114*, 3659.

2. Festa, R. A.; Thiele, D. J. *Curr. Biol.* **2011**, *21*, R877–83.

3. Koval, I. A.; Gamez, P.; Belle, C.; Selmeczi, K.; Reedijk, J. *Chem. Soc. Rev.* **2006**, *35*, 814.

4. Decker, H.; Schweikardt, T.; Tuczek, F. *Angew. Chem. Int. Ed.* **2006**, *45*, 4546.

5. van Holde, K. E.; Miller, K. I.; Decker, H. *J. Biol. Chem.* **2001**, *276*, 15563.

6. Rolff, M.; Schottenheim, J.; Decker, H.; Tuczek, F. *Chem. Soc. Rev.* **2011**, *40*, 4077.

7. Hamann, J. N.; Herzigkeit, B.; Jurgeleit, R.; Tuczek, F. *Coord. Chem. Rev.* **2017**, *334*, 54.

8. Solem, E.; Tuczek, F.; Decker, H. *Angew. Chem. Int. Ed.* **2016**, *55*, 2884.

9. Battaini, G.; Granata, A.; Monzani, E.; Gullotti, M.; Casella, L. *Adv. Inorg. Chem.* **2006**, *58*, 185.

10. a) Mirica, L. M.; Ottenwaelder, X.; Stack, T. D. P. *Chem. Rev.* **2004**, *104*, 1013. b) Lewis, E. A.; Tolman, W. B. *Chem. Rev.* **2004**, *104*, 1047. c) Que, L.; Tolman, W. B. *Nature.* **2008**, *455*, 333. d) Hatcher, L. Q.; Karlin, K. D. *J. Biol. Inorg. Chem.* **2004**, *9*, 669. e) Karlin, K. D.; Itoh, S.; Rokita, S. *Copper-oxygen chemistry*. Wiley-VCH, Hoboken, N. J., **2011**, f) Quist, D. A.; Diaz, D. E.; Liu, J. J.; Karlin, K. D. *J. Biol. Inorg. Chem.* **2017**, *22*, 253.

11. Elwell, C. E.; Gagnon, N. L.; Neisen, B. D.; Dhar, D.; Spaeth, A. D.; Yee, G. M.; Tolman, W. B. *Chem. Rev.* **2017**, *117*, 2059.

12. Than, R.; Feldmann, A. A.; Krebs, B. *Coord. Chem. Rev.* **1999**, *182*, 211.

13. Simándi, L. I.; Simándi, T. M.; May, Z.; Besenyei, G. *Coord. Chem. Rev.* **2003**, *245*, 85.

14. Gerdemann, C.; Eicken, C.; Krebs, B. *Acc. Chem. Res.* **2002**, *35*, 183.

15. Selmeczi, K.; Réglier, M.; Giorgi, M.; Speier, G. *Coord. Chem. Rev.* **2003**, *245*, 191.

16. Dey, S. K.; Mukherjee, A. *Coord. Chem. Rev.* **2016**, *310*, 80.

17. a) Osório, Renata, E. H. M. B.; Neves, A.; Camargo, T. P.; Mireski, S. L.; Bortoluzzi, A. J.; Castellano, E. E.; Haase, W.; Tomkowicz, Z. *Inorg. Chim. Acta* **2015**, *435*, 153. b) Banu, K. S.; Chattopadhyay, T.; Banerjee, A.; Bhattacharya, S.; Suresh, E.; Nethaji, M.; Zangrando, E.; Das, D. *Inorg. Chem.* **2008**, *47*, 7083. c) Banu, K. S.; Chattopadhyay, T.; Banerjee, A.; Bhattacharya, S.; Zangrando, E.; Das, D. *J. Mol. Catal. A: Chem.* **2009**, *310*, 34. d) Wegner, R.; Gottschaldt, M.; Görls, H.; Jäger, E.-G.; Klemm, D. *Chem. Eur. J.* **2001**, *7*, 2143. e) Smith, S. J.; Noble, C. J.; Palmer, R. C.; Hanson, G. R.; Schenk, G.; Gahan, L. R.; Riley, M. J. *J. Biol. Inorg. Chem.* **2008**, *13*, 499. f) Monzani, E.; Quinti, L.; Perotti, A.; Casella, L.; Gullotti, M.; Randaccio, L.; Geremia, S.; Nardin, G.; Faleschini, P.; Tabbì, G. *Inorg. Chem.* **1998**, *37*, 553. g) Garcia-Bosch, I.; Karlin, K. D. *Copper Peroxide Bioinorganic Chemistry: From Metalloenzymes to Bioinspired Synthetic Systems*, Wiley-VCH, Hoboken, N. J., **2014**.

18. Lo Presti, E.; Monzani, E.; Santagostini, L.; Casella, L. *Inorg. Chim. Acta* **2018**, *481*, 47.

19. Karlin, K. D.; Hayes, J. C.; Gultneh, Y.; Cruse, R. W.; McKown, J. W.; Hutchinson, J. P.; Zubieta, J. *J. Am. Chem. Soc.* **1984**, *106*, 2121.

20. Pidcock, E.; Obias, H. V.; Zhang, C. X.; Karlin, K. D.; Solomon, E. I. *J. Am. Chem. Soc.* **1998**, *120*, 7841.

21. Pidcock, E.; Obias, H. V.; Abe, M.; Liang, H.-C.; Karlin, K. D.; Solomon, E. I. *J. Am. Chem. Soc.* **1999**, *121*, 1299.

22. Sanyal, I.; Mahroof-Tahir, M.; Nasir, M. S.; Ghosh, P.; Cohen, B. I.; Gultneh, Y.; Cruse, R. W.; Farooq, A.; Karlin, K. D.; Liu, S. *et al. Inorg. Chem.* **1992**, *31*, 4322.

23. Mahapatra, S.; Halfen, J. A.; Wilkinson, E. C.; Pan, G.; Cramer, C. J.; Que, L.; Tolman, JR. W. B. *J. Am. Chem. Soc.* **1995**, *117*, 8865.

24. López, I.; Cao, R.; Quist, D. A.; Karlin, K. D.; Le Poul, N. *Chem. Eur. J.* **2017**, *23*, 18314.

25. Gennarini, F.; David, R.; López, I.; Le Mest, Y.; Réglier, M.; Belle, C.; Thibon-Pourret, A.; Jamet, H.; Le Poul, N. *Inorg. Chem.* **2017**, *56*, 7707.

26. Holland, P. L.; Rodgers, K. R.; Tolman, W. B. *Angew. Chem.* **1999**, *111*, 1210.

27. Becker, J.; Gupta, P.; Angersbach, F.; Tuczek, F.; Näther, C.; Holthausen, M. C.; Schindler, S. *Chem. Eur. J.* **2015**, *21*, 11735.

28. Hamann, J. N.; Rolff, M.; Tuczek, F. *Dalton trans.* **2015**, *44*, 3251.

29. Karlin, K. D.; Cohen, B. I.; Farooq, A.; Liu, S.; Zubieta, J. *Inorg. Chim. Acta* **1988**, *153*, 9.

30. Tabuchi, K.; Ertem, M. Z.; Sugimoto, H.; Kunishita, A.; Tano, T.; Fujieda, N.; Cramer, C. J.; Itoh, S. *Inorg. Chem.* **2011**, *50*, 1633.

31. Itoh, S.; Abe, T.; Morimoto, Y.; Sugimoto, H. *Inorg. Chim. Acta* **2018**, *481*, 38.

32. Kunishita, A.; Kubo, M.; Sugimoto, H.; Ogura, T.; Sato, K.; Takui, T.; Itoh, S. *J. Am. Chem. Soc.* **2009**, *131*, 2788.

33. Rolff, M.; Hamann, J. N.; Tuczek, F. *Angew. Chem. Int. Ed.* **2011**, *50*, 6924.

34. Mirica, L. M.; Vance, M.; Rudd, D. J.; Hedman, B.; Hodgson, K. O.; Solomon, E. I.; Stack, T. D. P. *Science* **2005**, *308*, 1890.

35. Bulkowski, J. E. **1985**, 4545937,

36. Réglier, M.; Jorand, C.; Waegell, B. *J. Chem. Soc. Chem. Commun.* **1990**, *107*, 1752.

37. Rolff, M.; Schottenheim, J.; Peters, G.; Tuczek, F. *Angew. Chem. Int. Ed.* **2010**, *49*, 6438.

38. Hoffmann, A.; Citek, C.; Binder, S.; Goos, A.; Rübhausen, M.; Troeppner, O.; Ivanović-Burmazović, I.; Wasinger, E. C.; Stack, T. D. P.; Herres-Pawlis, S. *Angew. Chem. Int. Ed.* **2013**, *52*, 5398.

39. Schottenheim, J.; Fateeva, N.; Thimm, W.; Krahmer, J.; Tuczek, F.; *Z. anorg. allg. Chem.* **2013**, *639*, 1491.

40. Hamann, J. N.; Tuczek, F. *Chem. Commun.* **2014**, *50*, 2298.

41. Schottenheim, J.; Gernert, C.; Herzigkeit, B.; Krahmer, J.; Tuczek, F.; *Eur. J. Inorg. Chem.* **2015**, 3501.

42. Hamann, J. N.; Schneider, R.; Tuczek, F. *J. Coord. Chem.* **2015**, *68*, 3259.

43. Wilfer, C.; Liebhäuser, P.; Erdmann, H.; Hoffmann, A.; Herres-Pawlis, S. *Eur. J. Inorg. Chem.* **2015**, *2015*, 494.

44. Wendt, F.; Näther, C.; Tuczek, F. *J. Biol. Inorg. Chem.* **2016**, *21*, 777.

45. Herzigkeit, B.; Flöser, B. M.; Engesser, T. A.; Näther, C.; Tuczek, F. *Eur. J. Inorg. Chem.* **2018**, *2018*, 3058.

46. Trammell, R.; See, Y. Y.; Herrmann, A. T.; Xie, N.; Díaz, D. E.; Siegler, M. A.; Baran, P. S.; Garcia-Bosch, I. *J. Org. Chem.* **2017**, *82*, 7887.

47. Op't Holt, B. T.; Vance, M. A.; Mirica, L. M.; Heppner, D. E.; Stack, T. D. P.; Solomon, E. I. *J. Am. Chem. Soc.* **2009**, *131*, 6421.

48. Esguerra, K. V. N.; Fall, Y.; Lumb, J.-P. *Angew. Chem. Int. Ed.* **2014**, *53*, 5877.

49. Esguerra, K. V. N.; Fall, Y.; Petitjean, L.; Lumb, J.-P. *J. Am. Chem. Soc.* **2014**, *136*, 7662.

50. Askari, M. S.; Rodríguez-Solano, L. A.; Proppe, A.; McAllister, B.; Lumb, J.-P.; Ottenwaelder, X. *Dalton trans.* **2015**, *44*, 12094.

51. Askari, M. S.; Esguerra, K. V. N.; Lumb, J.-P.; Ottenwaelder, X. *Inorg. Chem.* **2015**, *54*, 8665.

52. Xu, B.; Lumb, J.-P.; Arndtsen, B. A. *Angew. Chem. Int. Ed.* **2015**, *54*, 4208.

53. Huang, Z.; Kwon, O.; Esguerra, K. V. N.; Lumb, J.-P. *Tetrahedron* **2015**, *71*, 5871.

54. a) Capdevielle, P.; Maumy, M. *Tetrahedron Lett.* **1982**, *23*, 1573. b) Capdevielle, P.; Maumy, M. *Tetrahedron Lett.* **1982**, *23*, 1577.

55. Casella, L.; Gullotti, M.; Bartosek, M.; Pallanza, G.; Laurenti, E. *J. Chem. Soc. Chem. Commun.* **1991**, 1235.

56. Santagostini, L.; Gullotti, M.; Monzani, E.; Casella, L.; Dillinger, R.; Tuczek, F. *Chem. Eur. J.* **2000**, *6*, 519.

57. Battaini, G.; de Carolis, M.; Monzani, E.; Tuczek, F.; Casella, L. *Chem. Commun.* **2003**, 726.

58. Garcia-Bosch, I.; Cowley, R. E.; Díaz, D. E.; Peterson, R. L.; Solomon, E. I.; Karlin, K. D. *J. Am. Chem. Soc.* **2017**, *139*, 3186.

59. Garcia-Bosch, I.; Cowley, R. E.; Díaz, D. E.; Siegler, M. A.; Nam, W.; Solomon, E. I.; Karlin, K. D. *Chem. Eur. J.* **2016**, *22*, 5133.

60. Park, G. Y.; Qayyum, M. F.; Woertink, J.; Hodgson, K. O.; Hedman, B.; Narducci Sarjeant, A. A.; Solomon, E. I.; Karlin, K. D. *J. Am. Chem. Soc.* **2012**, *134*, 8513.

61. Paul, M.; Teubner, M.; Grimm-Lebsanft, B.; Golchert, C.; Meiners, Y.; Senft, L.; Keisers, K.; Liebhäuser, P.; Rösener, T.; Biebl, F. *et al. Chem. Eur. J.* **2020**, *26*, 7556.

62. Schneider, R.; Engesser, T. A.; Näther, C.; Krossing, I.; Tuczek, F. *Angew. Chem. Int. Ed.* **2022**, *61*, e202202562.

63. Herzigkeit, B.; Flöser, B. M.; Meißner, N. E.; Engesser, T. A.; Tuczek, F. *ChemCatChem* **2018**, *10*, 5402.

64. Canty, A. J.; Minchin, N. J. *Aust. J. Chem.* **1986**, *39*, 1063.

152 Koch *et al.*

65. Thies, S.; Bornholdt, C.; Köhler, F.; Sönnichsen, F. D.; Näther, C.; Tuczek, F.; Herges, R. *Chem. Eur. J.* **2010**, *16*, 10074.

66. a) Abboud, J.-L. M.; Foces-Foces, C.; Notario, R.; Trifonov, R. E.; Volovodenko, A. P.; Ostrovskii, V. A.; Alkorta, I.; Elguero, J. *Eur. J. Org. Chem.* **2001**, *2001*, 3013. b) Lõkov, M.; Tshepelevitsh, S.; Heering, A.; Plieger, P. G.; Vianello, R.; Leito, I. *Eur. J. Org. Chem.* **2017**, *2017*, 4475. c) Frey, P. A.; Hegeman, A. D. *Enzymatic reaction mechanisms*, Oxford University Press, Oxford, **2007**.

67. Buncel, E.; Joly, H. A.; Jones, J. R. *Can. J. Chem.* **1986**, *64*, 1240.

68. Gradert, C.; Krahmer, J.; Sönnichsen, F. D.; Näther, C.; Tuczek, F. *J. Organomet. Chem.* **2014**, *770*, 61.

69. Solomon, E. I.; Sundaram, U. M.; Machonkin, T. E. *Chem. Rev.* **1996**, *96*, 2563.

70. Herzigkeit, B.; Jurgeleit, R.; Flöser, B. M.; Meißner, N. E.; Engesser, T. A.; Näther, C.; Tuczek, F. *Eur. J. Inorg. Chem.* **2019**, *2019*, 2258.

71. Kwon, O.; Esguerra, K.; Glazerman, M.; Petitjean, L.; Xu, Y.; Ottenwaelder, X.; Lumb, J.-P. *Synlett* **2017**, *28*, 1548.

72. Nilges, M. J.; Swartz, H. M.; Riley, P. A. *J. Biol. Chem.* **1984**, 2446.

73. Bailey, S. I.; Ritchie, I. M.; Hong-Guang, Z. *Bioelectrochem. Bioenerg.* **1988**, *19*, 521.

74. Waite, J. H. *Anal. Biochem.* **1976**, *75*, 211.

75. Rouet-Mayer, M.-A.; Ralambosoa, J.; Philippon, J. *Phytochemistry* **1990**, *29*, 435.

76. Taylor, S. W.; Molinski, T. F.; Rzepecki, L. M.; Waite, J. H. *J. Nat. Prod.* **1991**, *54*, 918.

5 Electrochemistry and Spectroelectrochemistry of Copper-Oxygen-Relevant Species

Nicolas Le Poul

Laboratoire CEMCA, UMR CNRS 6521, Université de Bretagne Occidentale, 29238 Brest, France
lepoul@univ-brest.fr

I. Introduction

Redox reactions involving molecular oxygen (O_2) are ubiquitous in chemistry and biology.[1] For instance, the hydroxylation of dopamine into norepinephrine in the catecholamine synthetic pathway is catalyzed by the copper-based DβM glycoprotein (DβM = Dopamine β Monooxygenase). This redox enzyme, among others such as the particulate methane monooxygenase (pMMO)[2] or peptidylglycine α-hydroxylating monooxygenase (PHM),[3] is able to cleave the O–O bond of O_2 and to perform the incorporation of one oxygen atom into a C–H bond at a specific position. Alternatively, the four-electron reduction of dioxygen into water is carried out at the biological scale by oxidases such as laccases or cytochrome *c* oxidase, where dioxygen serves as a pure electron/proton acceptor, thus inhibiting the generation of partially reduced oxygen species (PROS).[4] For more than 40 years now, these enzymes have been a source of inspiration for chemists[5] aiming at developing efficient catalysts for synthetic chemistry, fuel cell devices, solar photoelectrochemical cells, or organic lithium–air batteries.[6] Since dioxygen is kinetically inert toward organic substrates, its activation is classically carried out by reduced metals (M = Fe, Cu, Mn) in active synthetic and biological systems.[1a,7] For instance, the reaction of one or several Cu(I) centers with O_2 results in the formation of mono- or polynuclear metal–oxygen adducts such as copper superoxide, peroxides, or hydroperoxides (Figure 1). These transient species are particularly reactive and can evolve under certain conditions (presence of acid and/or reductants) toward ultra-reactive high-valent species.[1b,5] The latter have attracted much attention these last years because of their potential use as catalysts for hydrogen-atom

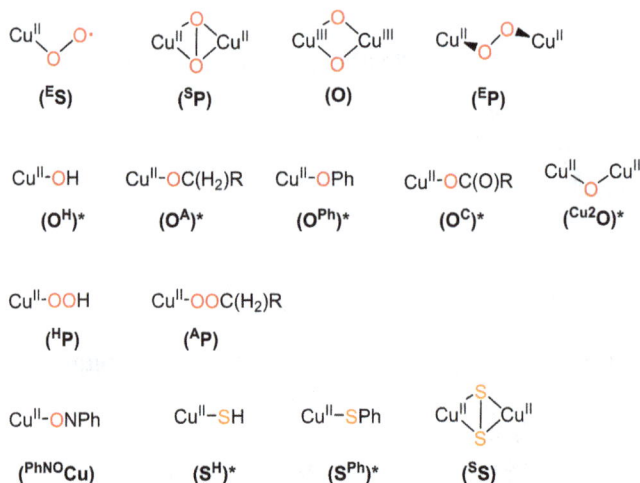

Figure 1. Schematic representation of the different copper–oxygen relevant species including precursors and sulfur derivatives investigated by electrochemistry and considered in this chapter. Abbreviations: ^{E}S: *end-on* superoxo; ^{S}P: *side-on* peroxy; O: bis(μ-oxo); ^{E}P: *end-on* peroxy; O^H: hydroxo; O^A: alkoxo; O^{Ph}: phenoxy; O^C: carboxylato; ^{Cu2}O: μ-oxo; ^{H}P: hydroperoxy; ^{A}P: alkoperoxy; ^{PhNO}Cu: nitrosoarene; S^H: hydrosulfido; S^{Ph}: thiophenolato; ^{S}S: *side-on* disulfido. *Precursors for reactive species.

abstraction (HAA) of strong C–H bonds of alkanes (BDFE > 90 kcal mol^{-1}, BDFE = Bond Dissociation Free Energy).[8] Significant progress has been recently made on the elucidation of the different mechanistic pathways, which can occur for HAA reactions involving $Cu_n:O_2$ species. In particular, the concerted or stepwise features of the coupled electron–proton reactions have been thoroughly investigated in both thermodynamics and kinetics terms.[9] This has often, but not always, required the determination of redox potential values of the copper–oxygen adducts by either chemical or electrochemical means.

Independently, many copper systems have also been studied as potential electrocatalysts for the oxygen reduction reaction (ORR) in fuel cells[10,11] and oxygen evolution reaction (OER) from water oxidation.[12] Most of reported studies have been carried out by using voltammetric methods, with the aim of quantifying the electrode kinetics from electrochemical analysis (Tafel slope, E/pH variation). More rarely, key active copper–oxygen species have been characterized by decoupling electron and proton transfers in presence of organic solvents.[13]

As a matter of fact, electrochemical and spectroelectrochemical characterizations of $Cu_n:O_2$ species have been seldom reported, although they have been well investigated by UV–vis or resonance Raman spectroscopies.[5,14] One main reason is that these species are often unstable in the conditions of experimental measurement (room temperature). As it will be shown in the following sections throughout chosen examples, several strategies have been proposed to overcome this issue: (i) modification of the ligand topology by using a pre-organized framework to trap O_2, or by inserting H-donors to interact with O atoms; (ii) *in situ* electrochemical reduction of stable Cu(II) species, which can further react with O_2, or electrochemical oxidation of stable Cu(II) precursors such as such as copper(II) hydroxo, alkoxo, or carboxylato cores (Figure 1) into reactive oxidized species; and (iii) decrease of the temperature of measurement to increase the lifetime of the copper–oxygen adduct. These three main strategies have been successfully applied (sometimes combined) and have allowed electrochemical characterization of several species. An alternative approach based on the reaction of copper–oxygen adducts with chemical reductants/oxidants has also been developed these last years.[15]

Nevertheless, this indirect procedure yields only approximated values of thermodynamic and kinetic parameters.

From this statement, the purpose of this chapter is to gather electrochemical data existing on copper–oxygen relevant species (Figure 1) and to demonstrate how electrochemistry and spectroelectrochemistry can bring useful information and help for getting mechanistic insights on dioxygen activation pathways. The chapter has been divided into four sections according to the different families of complexes shown in Figure 1. The first section describes the electrochemistry of superoxides and peroxides species formed upon reaction of Cu(I) complexes with dioxygen. For most of these complexes, low-temperature was necessary to allow full stabilization and characterization by electrochemical and spectroelectrochemical techniques. The second section is focused on the redox properties of mono/dicopper(II) hydroxides, alkoxides, and carboxylates species, which are precursors for reactive Cu(III)–oxygen adducts. The third section aims at describing the electrochemistry of Cu(II) hydroperoxides and alkoperoxides complexes, which can themselves act as reactive species toward hydrogenated substrates, or lead upon O–O bond cleavage to high-valent oxyl adducts. The fourth section discusses about the redox chemistry of copper(II) phenoxides complexes, which yield reactive phenoxyl metal species upon oxidation. At last, the fifth section describes the recent electrochemical studies of sulfides and nitrosoarene analogues of copper–oxygen adducts. For each section, the most striking examples have been chosen and illustrated.

II. Electrochemistry of Peroxide and Superoxide Copper–Oxygen Species

As shown in Figure 2, mono or dinuclear superoxide and peroxide copper species can be generated from the reaction of Cu(I) complexes with dioxygen. For a great majority of cases, these copper–oxygen adducts are highly unstable and require the use of low temperature to stabilize them, making their experimental characterization quite difficult. Since Cu(I) complexes are highly O_2-sensitive, they can also be prepared *in situ* from their Cu(II) analogues by using either chemical or electrochemical reduction under inert conditions. Once stabilized at low temperature, Cu(I) can be reacted with O_2 to generate Cu(II) superoxide or peroxide species.

Mononuclear species

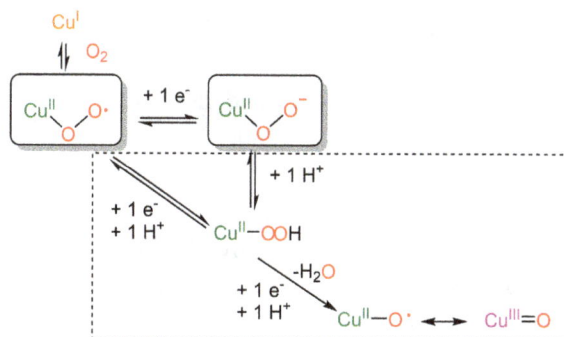

Dinuclear species

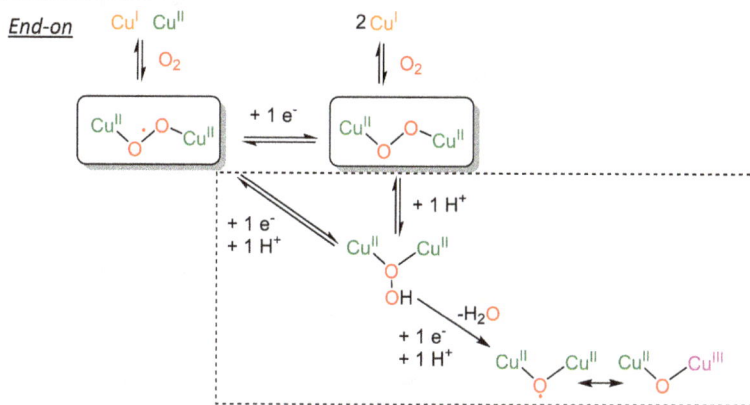

Side-on

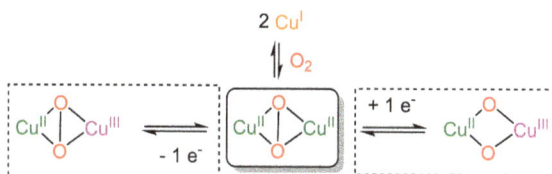

Figure 2. Electron and proton–electron transfer reactions for mono and dinuclear superoxo and peroxo copper species. Plain-line rectangles designate the species of interest, here superoxo and peroxo Cu(II) cores. Dotted-line rectangles indicate reactions that have not yet been evidenced by electrochemical methods.

As shown in Figure 2, monoelectronic reduction of Cu(II)-superoxides leads to Cu(II)-peroxides. In presence of proton source, this monoreduction produces Cu(II) hydroperoxide species, which may further undergo O–O bond cleavage through further e^-, H^+ transfer. The resulting Cu(II)-oxyl species have been postulated as being strong oxidant for HAA reactions.[16] Figure 2 also displays the redox chemistry starting from *end-on* or *side-on* dinuclear peroxo species. In particular, monoelectronic reduction of *side-on* dicopper(II) peroxides may lead to O–O bond cleavage and formation of mixed-valent Cu(II)–Cu(III) bis(μ-oxo) adducts. The latter have been suggested to be key species in the oxidation of methane in pMMOs.[17]

First investigations on redox properties of peroxide/superoxide copper–oxygen species were initiated in 1992 by Karlin and co-workers for three different *side-on* peroxo dicopper(II) complexes **1–3** bearing two linked bis-pyridyl(ethylamine) moieties (Figure 3).[18] These adducts were prepared at low temperature (193 K) in dichloromethane from the reaction of bis-Cu(I) complexes with dioxygen. UV–visible spectroscopic analyses supported by EXAFS studies of these adducts evidenced the formation of bent-butterfly peroxo dicopper(II) species resulting from an

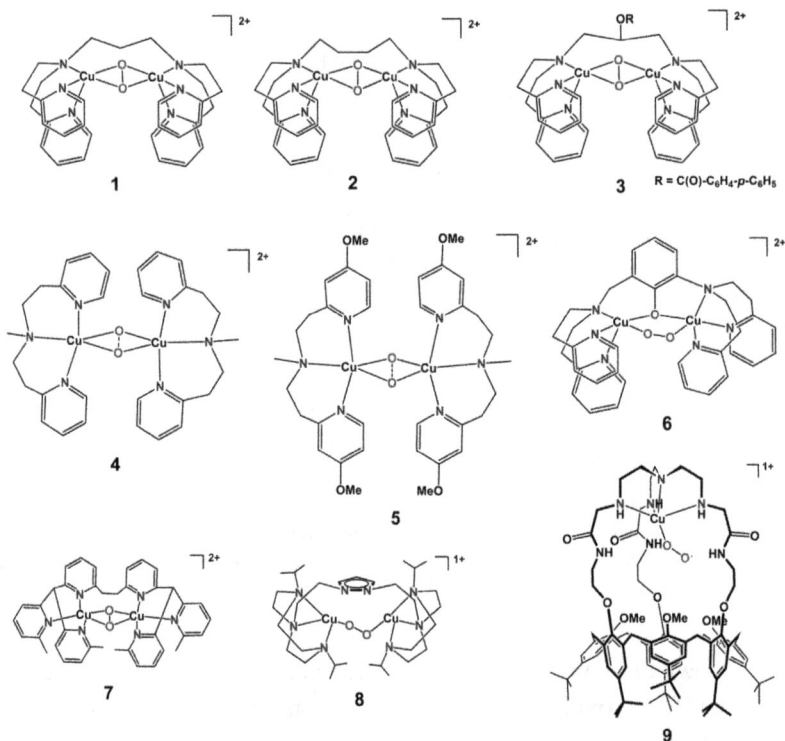

Figure 3.   Schematic representation of peroxo and superoxo complexes **1–9**.

Table 1. Electrochemical data for *side-on* peroxo (^{S}P), *end-on* peroxo (^{E}P), bis(μ-oxo) (O), and *end-on* superoxo (^{E}S) complexes **1–9** (red: reduction; ox: oxidation).

Complex	$E_{1/2}$ / V vs. Fc$^+$/Fc	Conditions	Adduct	Ref.
1	−1.77 (red)[a,b]	193 K, CH$_2$Cl$_2$/NBu$_4$PF$_6$	^{S}P	[18]
2	−1.57 (red)[a,b]	193 K, CH$_2$Cl$_2$/NBu$_4$PF$_6$	^{S}P	[18]
3	−1.87 (red)[a,b]	193 K, CH$_2$Cl$_2$/NBu$_4$PF$_6$	^{S}P	[18]
4	0.08 (red)[c]	195 K, CH$_2$Cl$_2$/NBu$_4$(B(C$_6$F$_5$)$_4$)	^{S}P/O	[20]
5	−0.03 (red)[c]	195 K, CH$_2$Cl$_2$/NBu$_4$(B(C$_6$F$_5$)$_4$)	^{S}P/O	[20]
6	−0.36 (ox)	193 K, CH$_2$Cl$_2$/NBu$_4$ClO$_4$	^{E}P	[21]
7	−0.75 (red)[a]	293 K, CH$_2$Cl$_2$/NBu$_4$PF$_6$	^{S}P	[19]
8	−0.59 (ox)[d]	273 K, CH$_3$CN/NBu$_4$PF$_6$	^{E}P	[9c]
9	< −0.90 (red)	213 K, Acetone/NBu$_4$PF$_6$	^{E}S	[22]

[a]Irreversible peak; [b]Recalculated *versus* Fc$^+$/Fc by taking E^0(Fc$^+$/Fc) = 0.67 V *versus* Ag/AgCl in this solvent; [c]Recalculated *versus* Fc$^+$/Fc by taking E^0(Fc$^+$/Fc) = 0.47 V *versus* SCE in this solvent; [d]Recalculated *versus* Fc$^+$/Fc by taking E^0(Fc$^+$/Fc) = 0.08 V *versus* Ag/AgNO$_3$ in this solvent.

intramolecular O$_2$ binding. Low-temperature CV of the peroxo complexes performed under argon displayed an irreversible reduction peak varying between −1.57 V and −1.87 V *versus* Fc$^+$/Fc (Table 1), a potential value close to that of free dioxygen.[19] A comparative voltammetric analysis suggested that the process was likely monoelectronic. No oxidation peak could be detected until ca. 1.4 V *versus* Fc$^+$/Fc, hence demonstrating the difficulty to generate any superoxo bis-Cu(II) species.

Karlin and co-workers reported a few years later a voltammetric study of peroxo complexes **4** and **5** (Figure 3).[20] These adducts, generated by reaction of mononuclear cuprous complexes [Cu^I(Me^{R}PY2)]$^+$ with dioxygen, were previously shown to be efficient catalysts for HAA reactions for a wide range of hydrogenated substrates. Low-temperature (195 K) CVs were carried out in dichloromethane by using NBu$_4$(B(C$_6$F$_5$)$_4$) as supporting electrolyte. Noteworthy, both complexes **4** and **5** displayed a quasi-reversible system upon reduction at +0.08 V and −0.03 V *versus* Fc$^+$/Fc (Figure 4 and Table 1), respectively, which was ascribed to the formation of transient mixed-valent Cu$_2^{III/II}$ bis(μ-oxo) species. From these thermodynamic data and with the support of complementary studies, the authors concluded that HAA of substrates by complexes **4** and **5** was likely occurring through a concerted electron–proton transfer reaction (Figure 4).

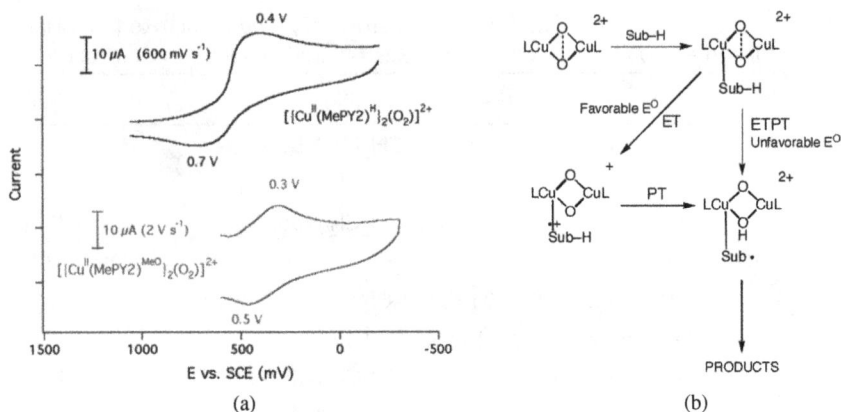

Figure 4. (a) CVs of complexes **4** (top) and **5** (bottom) (T = 195 K) in CH_2Cl_2/ $NBu_4(B(C_6F_5)_4)$. (b) Concerted and stepwise mechanistic pathways considered for HAA by these complexes. Adapted with permission from Ref. [20]. Copyright (2005) American Chemical Society.

In 2017, the groups of Karlin and Le Poul on one side, as well as that of Meyer on the other side reported the electrochemical properties of *end-on* peroxo complexes **6** and **8**.[9c,21] For the complex **8** (Figure 3),[9c] cyclic voltammetry analysis in acetonitrile at 0°C was sufficient to demonstrate the reversibility of the peroxide/superoxide reaction at −0.59 V *versus* Fc⁺/Fc (Table 1). This half-wave value was further utilized to calculate the BDFE(O-H) (72 kcal mol⁻¹) of the corresponding hydroperoxo species knowing the pKa for the peroxo/hydroperoxo couple (22.2 in MeCN). For the peroxo complex **6** (Figure 3), Lopez *et al.* investigated several aspects of the peroxide/superoxide reaction upon oxidation of the complex at 193 K.[21] A reversible monoelectronic system (Figure 5) was detected at a slightly higher potential (−0.36 V *vs.* Fc⁺/Fc) than for complex **8** (Table 1). Electrochemical impedance spectroscopy allowed the determination of the electron transfer kinetics for this process. The relatively low inner-sphere reorganizational energy (0.11 eV) was ascribed to an electron-transfer process occurring exclusively on the O_2 core. Noteworthy, *in situ* UV–vis characterization of the superoxide was carried out at 193 K by a cryo-spectroelectrochemical setup (Figure 5). Reduction of the superoxide allowed the back formation of the complex **6**.

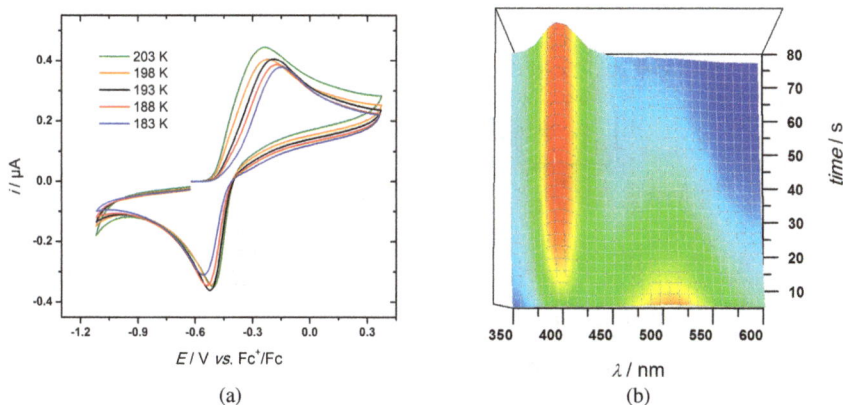

Figure 5. (a) CVs of complex **6** in CH_2Cl_2/NBu_4ClO_4 for $183 < T < 203$ K. (b) Low-temperature (193 K) UV–vis *in situ* spectroelectrochemical monitoring of the peroxide oxidation (absorbance at 530 nm) leading to the formation of the superoxide complex (absorbance at 406 nm). Adapted from Ref. [21] with permission from John Wiley and Sons.

Electrochemical studies of the room-temperature stable *side-on* peroxo complex **7** was performed by Le Poul and co-workers in 2018.[19] The complex exhibited an irreversible two-electron reduction peak at −0.75 V versus Fc^+/Fc, which was assigned to the O–O bond cleavage. Spectroelectrochemical measurements showed a complete disappearance of the peroxo characteristic band at 365 nm upon reduction, while chemical analysis after exhaustive electrolysis indicated no formation of hydrogen peroxide but merely a dicopper(II) bis(μ-OH) species. More recently, de Leener *et al.* studied the reaction of a calix[6]trenamide Cu(I) complex with dioxygen.[22] Low-temperature spectroelectrochemistry in acetone showed the formation of a transient superoxo species (complex **9**, Figure 3) by *in situ* reduction of the Cu(II) complex. Although no further reduction could be detected by CV, the data suggested that reduction of the superoxide could only occur at lower potential values than −0.90 V *versus* Fc^+/Fc. Noteworthy, room temperature investigations demonstrated that the copper superoxo could perform self-oxidation of the ligand (insertion of an oxygen atom) despite its low oxidation potential value.

Mononuclear species

R=H, Alkyl, OCR'

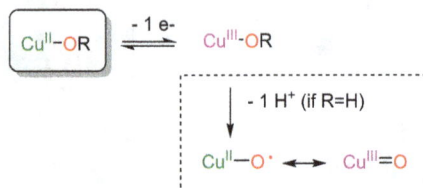

Dinuclear species

R=H, Alkyl, OCR'

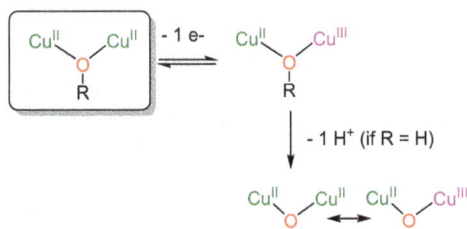

Figure 6. Electron and proton–electron transfer reactions for mono and dinuclear hydroxo, alkoxo, and carbonato copper species. Plain-line rectangles designate the species of interest, here hydroxo, alkoxo, and carbonato Cu(II) cores. The dotted-line rectangle indicates a reaction that has not yet been evidenced by electrochemical methods.

III. Electrochemistry of Copper(II) Hydroxo, Alkoxo, Carboxylato, and Oxo Precursors

Alternatively to the classical Cu(I)/O$_2$ approach described above, a quite recent strategy based on the electrochemical oxidation of Cu(II) hydroxo, alkoxo, or carboxylato complexes has been reported. The principle is to generate a Cu(III) oxygen species from a stable Cu(II) precursor, which can be easily handled at room temperature. This simple approach allows direct generation of high-valent copper oxygen adducts. In the case of hydroxo complexes, further deprotonation of the Cu(III)-OH species may yield Cu(III)-oxo core (Figure 6).

This strategy was first initiated by Tolman and co-workers in 2014. As a first example, a dicopper(II) mono-μ-hydroxo complex bearing a

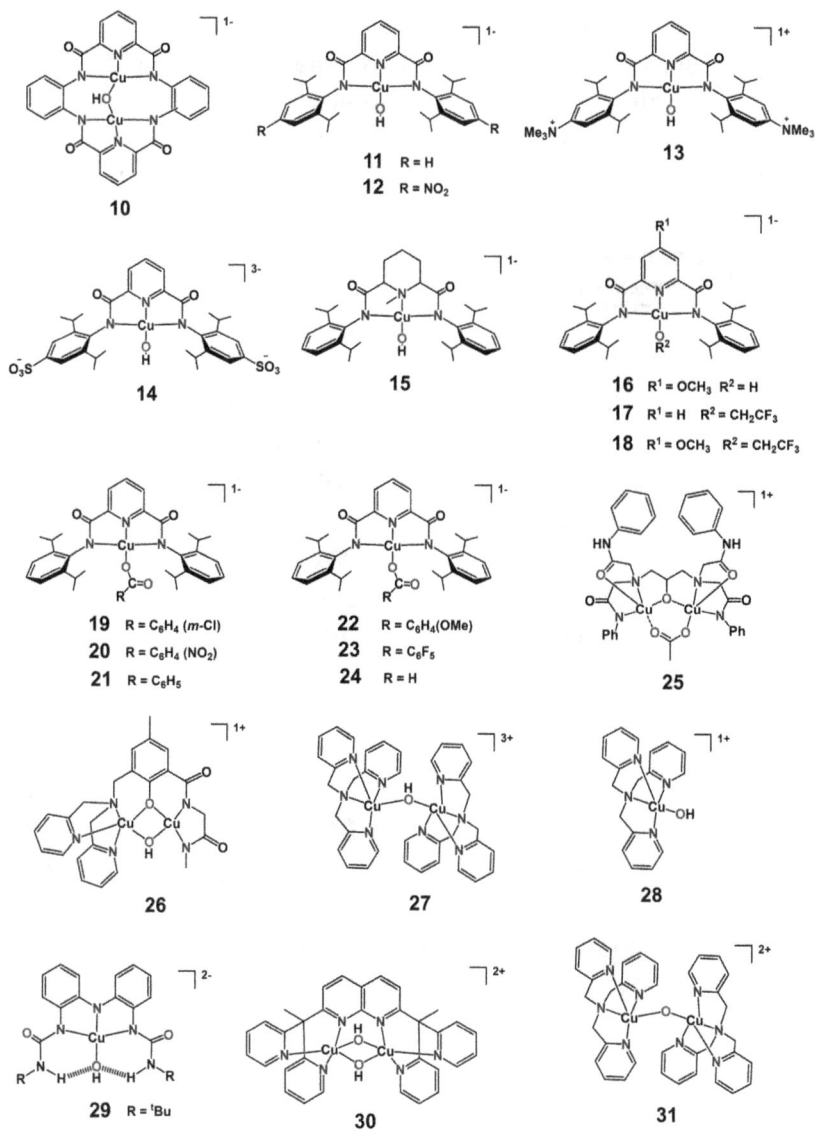

Figure 7. Schematic representation of hydroxo, alkoxo, carboxylate, and oxo precursors **10–31**.

macrocyclic ligand with two pyridine(dicarboxamide) moieties was reported (complex **10**, Figure 7).[23] This anionic and strongly donating ligand was specifically chosen in order to stabilize the high-valent Cu(III) states of the complex upon oxidation. Room temperature voltammetry experiments in wet DMF exhibited two reversible systems at 0.18 and

Table 2. Electrochemical data for hydroxo (O^H), alkoxo (O^A), carboxylato (O^C), and μ-oxo (^{Cu2}O) complexes **10–31** (red: reduction; ox: oxidation).

Complex	$E_{1/2}$ / V vs. Fc$^+$/Fc	Conditions	Adduct	Ref.
10	0.18 (ox1); 0.47 (ox2)	293 K, DMF/NBu$_4$PF$_6$	O^H	[23]
11	−0.17 (ox)	293 K, Acetone or THF/ NBu$_4$PF$_6$	O^H	[8a,24]
12	0.12 (ox)	293 K, DFB/NBu$_4$PF$_6$	O^H	[15b]
13	0.14 (ox)	293 K, DFB/NBu$_4$BAr$^F_4$	O^H	[25]
14	−0.13 (ox)	293 K, DFB/NBu$_4$PF$_6$	O^H	[25]
15	−0.26 (ox)	293 K, DFB/NBu$_4$PF$_6$	O^H	[15b]
16	−0.20 (ox)	293 K, THF/NBu$_4$PF$_6$	O^H	[24b]
17	0.04 (ox)	293 K, THF/NBu$_4$PF$_6$	O^A	[24b]
18	−0.02 (ox)	293 K, THF/NBu$_4$PF$_6$	O^A	[24b]
19	0.23 (ox)	293 K, THF/NBu$_4$PF$_6$	O^C	[26]
20	0.24 (ox)	293 K, THF/NBu$_4$PF$_6$	O^C	[26]
21	0.17 (ox)	293 K, THF/NBu$_4$PF$_6$	O^C	[26]
22	0.15 (ox)	293 K, THF/NBu$_4$PF$_6$	O^C	[26]
23	0.30 (ox)	293 K, THF/NBu$_4$PF$_6$	O^C	[26]
24	0.15 (ox)	293 K, THF/NBu$_4$PF$_6$	O^C	[26]
25	0.63 (ox)[a]	293 K, DMF/NBu$_4$PF$_6$	O^C	[27]
26	0.12 (ox1); 0.64 (ox2)[a]	213 K, DMF/NBu$_4$ClO$_4$	O^H	[28]
27	−0.48 (red);1.87 (ox)	243 K, MeCN/NBu$_4$ClO$_4$	O^H	[9d,e]
28	0.54 (ox)	243 K, MeCN/NBu$_4$ClO$_4$	O^H	[9d]
29	−1.05 (ox1);−0.28 (ox2)	293 K, DMF/NBu$_4$PF$_6$	O^H	[29]
30	1.26 (ox)	293 K, MeCN/NBu$_4$PF$_6$	O^H	[30]
31	0.76 (ox1); 1.35 (ox2)	243 K, MeCN/NBu$_4$ClO$_4$	^{Cu2}O	[9d]

[a]Irreversible peak.

0.47 V *versus* Fc$^+$/Fc, which were ascribed to the generation of Cu$_2^{(III)/(II)}$ and Cu$_2^{(III)/(III)}$ μ-hydroxo species (Table 2). From these values, low-T chemical oxidation with appropriate oxidants led to the UV–vis and EPR spectroscopic characterization of the mono- and bis-oxidized species. With the support of TD-DFT, the authors suggested that the bridging ligand remained as a hydroxo group (rather than an oxo one), and that oxidation steps occurred effectively on the metal ions.

The same group then carried out similar studies with a series of mono-nuclear Cu(II)-hydroxo complexes bearing one pyridine(dicarboxamide) moiety (complexes **11–16**, Figure 6). The subtle modification of the ligand by various substituting groups allowed fine tuning of the Cu(II)–Cu(III) oxidation potential from –0.26 V to 0.14 V *versus* Fc$^+$/Fc according to their room-temperature voltammetric measurements (Table 2). Interestingly for complex **11**, the electrochemical oxidation was found to be reversible at moderate scan rate in acetone[24a] while it was irreversible in THF.[8a] In complement, partial reversibility was recovered in deuterated THF at 0.2 V s^{-1} (Figure 8),[8a] hence evidencing the reaction of the electrochemically generated Cu(III) species with THF. This effect was further analyzed in the frame of a HAA reaction for which THF acts as substrate (BDFE = 92 kcal mol^{-1}) through the formation of a Cu(III)-OH-THF adduct (Figure 8).

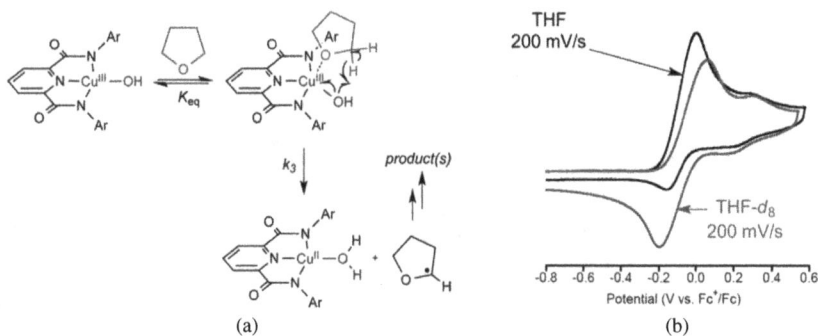

Figure 8. (a) Mechanistic pathway for THF oxidation mediated by the mono-oxidized Cu(III)-OH form of complex **11**. (b) Room-temperature CV at 0.2 V s^{-1} of complex **11** in THF/NBu$_4$PF$_6$ (black) and THF-d$_8$/NBu$_4$PF$_6$ (red). Adapted with permission from Ref. [8a]. Copyright (2015) American Chemical Society.

More recently, various alkoxo (complexes **17–18**, Figure 7) and carboxylato (complexes **19–24**, Figure 7) Cu(II) adducts of this series were prepared and studied by CV. For the carboxylate precursors, the oxidation potential values were found to be more positive than for the hydroxo analogues **11** and **17**, as expected from their lower basicity. Furthermore, the donating/withdrawing properties of the carboxylate ligand were shown to impact significantly the redox potential value (150 mV variation over the series; see Table 2), and consequently modify the HAA properties. According to the authors, the experimental data evidenced an asynchronous PCET oxidative mechanism for the reaction of the Cu(III)-carboxylate species with tris-tertbutylphenol.

Other examples of hydroxo, alkoxo, or carboxylato complexes bearing a different ligand backbone than the pyridine(dicarboxamide) moiety have also been reported for the last five years (see complexes **25–31**, Table 2, and Figure 7). For instance, Thibon-Pourret et al. synthesized a dicopper(II) hydroxide-phenoxide complex comprising an unsymmetrical ligand with a N,N-bis(2-pyridyl)methylamine moiety on one side and a dianionic bis-amide on the other side (complex **26**, Figure 7).[28] Electrochemical studies of the complex at low temperature in DMF allowed the reversible oxidation according to a monoelectronic process (Figure 9). *In situ*

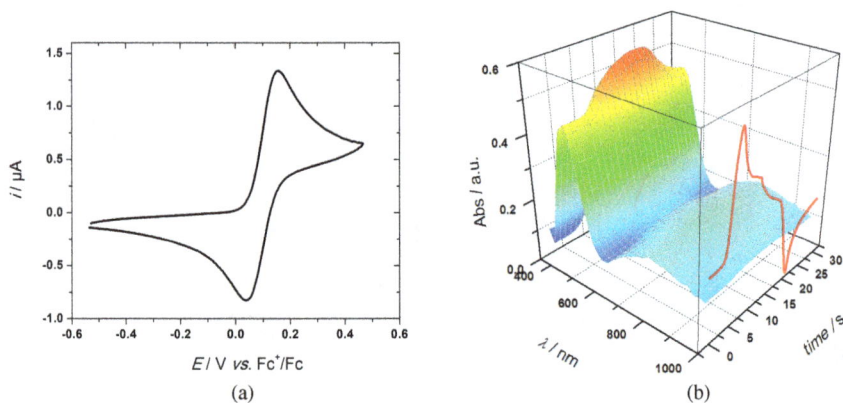

Figure 9. (a) Low-temperature (213 K) CV at $v = 0.05$ V s^{-1} of complex **26** in DMF/NBu$_4$ClO$_4$. (b) UV–vis *in situ* spectroelectrochemical monitoring of the oxidation at 213 K leading to the formation of the mixed-valent Cu$^{III/II}$ hydroxide species (red curve: *i vs. t* from CV plots). Adapted with permission from Ref. [28]. Copyright (2018) American Chemical Society.

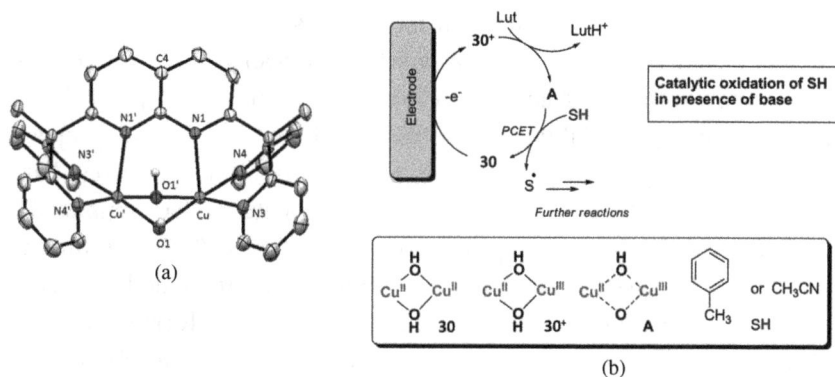

Figure 10. (a) X-ray structure of complex **30**. (b) Electrocatalytic oxidation of SH by the complex **30⁺** in presence of 2,6-lutidine (Lut). Adapted from Ref. [31] with permission of the Royal Society of Chemistry.

time-resolved UV–vis spectroelectrochemistry at $T = 213$ K showed the formation of a strong absorption band at 386 nm, which was further ascribed to a phenoxido-to-Cu(III) LMCT transition for the mixed-valent $Cu_2^{III/II}$ $\mu(O(H))$ species (Figure 9). The same research group published two papers on the oxidation of a bis-$\mu(OH)$ dicopper(II) complex bearing a naphthyridine-bridging ligand (complex **30**, Figure 7).[30–31] Noteworthy, mono-oxidation of this complex at 1.2 V *versus* Fc⁺/Fc was shown to yield a $Cu_2^{III/II}$ bis-$\mu(OH)$ adduct at room temperature from UV-vis-NIR spectroelectrochemical characterization.[30] Complementary works have demonstrated that the latter exhibited electrocatalytic HAA oxidative properties toward acetonitrile in presence of base (2,6-lutidine) through the formation of a putative $Cu_2^{III/II}$ $\mu(O)$ $\mu(OH)$ species (Figure 10).[31]

Starting from a dicopper(II) μ-hydroxo complex **27** (Figure 7) based on the TMPA ligand (TMPA = tris(methylpyridine)amine), Kieber-Emmons and co-workers investigated either the mono-reduction[9e] or mono-oxidation[9d] of this complex in order to generate $Cu^{II/I}$ and $Cu^{III/II}$ mixed-valent species, respectively. On one hand, low-T (243 K) CV in reduction yielded a reversible system at −0.48 V *versus* Fc⁺/Fc, which was assigned to a metal-centered process. From this value and knowing the pKa (24.3) for the dicopper(II) hydroxo/oxo couple (complexes **27** and **31**, Figure 7), the BDFE(OH) of the mixed-valent $Cu^{II/I}$ hydroxide species was found to be equal to 74.9 kcal mol^{-1}. This value indicated that the μ-oxo dicopper(II) complex **31** was a relatively poor oxidant for H-atom abstraction. On the other hand, irreversible oxidation of complex **27** was detected at high potential in dry and cooled (243 K) acetonitrile by square wave voltammetry (SWV) (1.87 V *vs.* Fc$^+$/Fc). Addition of a strong base (in excess) allowed the generation of the unprotonated μ-oxo complex **31**, which displayed an oxidation peak at 0.76 V *versus* Fc$^+$/Fc. According to the authors, the system was reversible in the SWV timescale hence suggesting the formation of the mono-oxidized oxo species, formally a Cu^{III}-O-Cu^{II} adduct. Further oxidation of this transient species was observed at 1.35 V *versus* Fc$^+$/Fc by SWV, which was assigned to the dicopper(III) μ(oxo) adduct. With the support of DFT calculations, and on the basis of determined pKa values, the authors concluded that both mono and bis-oxidized oxo adducts were strong oxidants for HAA (103.4 and 91.7 kcal mol^{-1}, respectively). Noteworthy, the better oxidative properties for the mono-oxidized were ascribed to the linear feature of the Cu-O-Cu core, which favored strong superexchange between Cu centers through the central oxygen and was merely considered as a dicopper(II)-oxyl adduct.

Alternatively, Garcia-Bosch and co-workers carried out room-temperature voltammetric studies of the mononuclear copper(II)-hydroxide complex **29** bearing a redox-active ligand comprising a tridentateamine and two ureanyl H-bonds donors (Figure 7).[29] Two reversible monoelectronic oxidation waves were detected at –1.05 V and –0.28 V *versus* Fc$^+$/Fc (Table 2). Noteworthy, UV–vis, EPR, and XAS spectroscopic methods using chemical oxidants evidenced that the electron transfer processes were essentially located on the ligand backbone and not on the metal. While the generated semi-quinone Cu(II) hydroxide species was found to be a moderate H-abstractor, the bis-oxidized quinone Cu(II) hydroxo complex exhibited a significant HAA 2e$^-$/2H$^+$ activity according to two sequential steps.

IV. Electrochemistry of Copper Hydroperoxide and Alkoxoperoxide Species

Copper(II) hydroperoxide and alkoxoperoxide species have recently drawn more attention because of the potential implication of either Cu(II) or Cu(III) hydroperoxide adducts in PCET processes occurring in

Mononuclear species

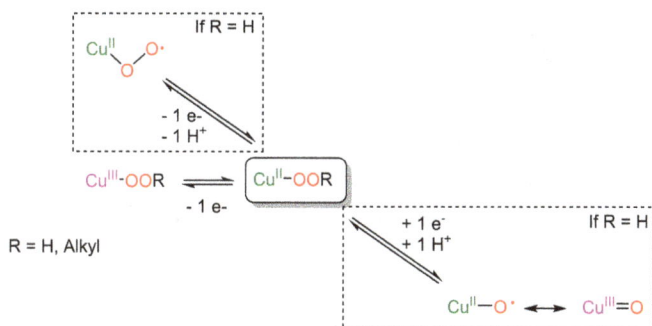

Dinuclear species

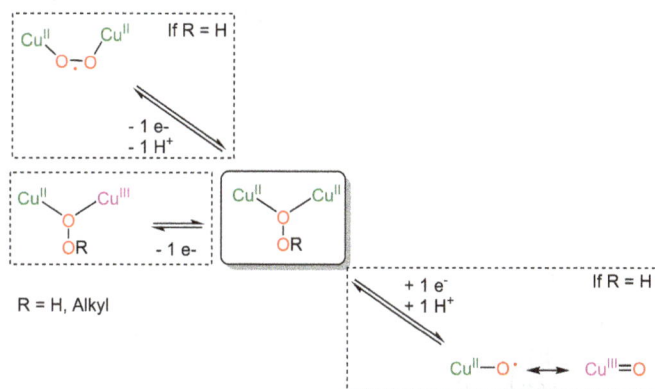

Figure 11. Electron and proton–electron transfer reactions for mono and dinuclear hydroperoxo or alkoxoperoxo copper species. Plain-line rectangles designate the species of interest, here hydroperoxo and alkylperoxo Cu(II) cores. Dotted-line rectangles indicate reactions that have not yet been evidenced by electrochemical methods.

monooxygenases such as PHM. As shown in Figure 11, oxidation of a Cu(II)-OOR core occurs on the metal center and leads preferentially to a Cu(III)-OOR species. The latter can perform oxidation reactions or potentially release OOR[•] by homolytic Cu–O bond cleavage.[32] Accordingly, 1e[−], 1H[+] reduction of Cu(II)-OOH species may lead to high-valent Cu(III)-oxo adducts, whereas 1e[−], 1H[+] oxidation may yield superoxo Cu(II) cores (Figure 11). Nevertheless, none of these two reactions has ever been reported so far by electrochemical approaches.

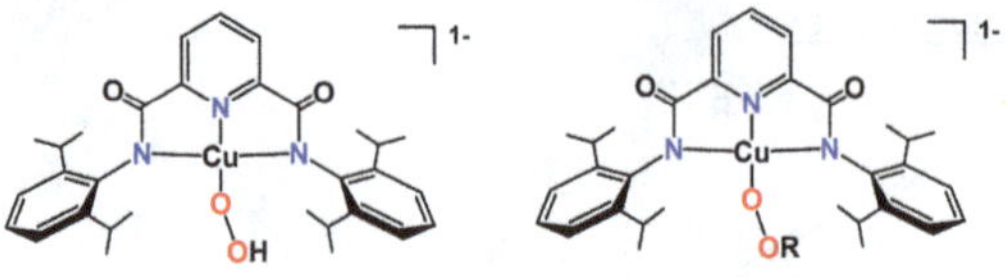

Figure 12. Schematic representation of hydroperoxide and alkylperoxide complexes **32–34**.

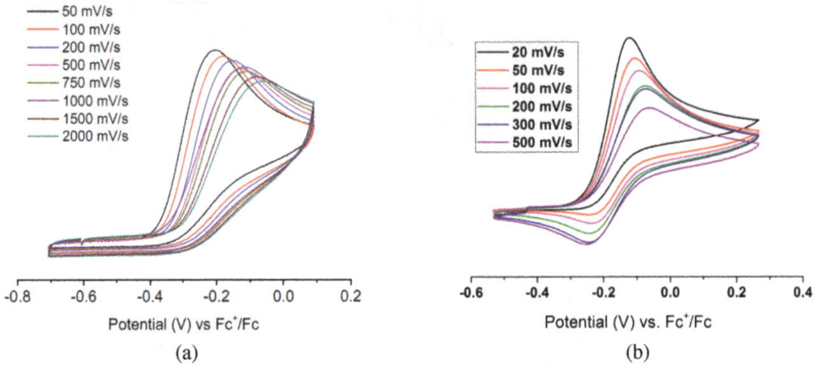

Figure 13. Normalized-scan-rate CVs of (a) complex **32** in DFB/NBu$_4$PF$_6$ at 273 K, and (b) complex **34** in THF/NBu$_4$PF$_6$ at 293 K. Adapted with permission from Refs. [9b] and [32]. Copyright (2017 and 2019) American Chemical Society.

Indeed, while many examples have been described on the electrochemistry of Cu(II)-hydroxide complexes, these for Cu(II) alkylperoxo and hydroperoxo remain scarce. Only three studies have been reported so far, all by the group of Tolman for mononuclear complexes (see complexes **32–34**, Figure 12).[9b,32] One main reason is that Cu(II)-OOH species are themselves reactive species toward hydrogenated substrates, as peroxide and superoxide copper complexes, and thus require low-temperature conditions to be stabilized and characterized. Moreover, their preparation based on the reaction of a Cu(II) hydroxo complex with hydrogen peroxide in presence of a base, is usually not quantitative. This renders the *in situ* electrochemical characterization not feasible.

Table 3. Electrochemical data for hydroperoxide (^{H}P) and alkylperoxide (^{A}P) complexes **32–34** (red: reduction; ox: oxidation).

Complex	$E_{1/2}$ / V vs. Fc$^+$/Fc	Conditions	Adduct	Ref.
32	−0.21 (ox)^a	273 K, DFB/NBu$_4$PF$_6$	^{H}P	[9b]
33	−0.20 (ox)^b	293 K, THF/NBu$_4$PF$_6$	^{A}P	[32]
34	−0.15 (ox)^c	293 K, THF/NBu$_4$PF$_6$	^{A}P	[32]

$^a E_{1/2}$ value assumed from CV simulations. Irreversible peak found at −0.18 V versus Fc$^+$/Fc at $v = 0.1$ V s^{-1} Irreversible peak; $^b E_{1/2}$ value taken at $v = 0.5$ V s^{-1}; $^c E_{1/2}$ value taken at $v = 0.5$ V s^{-1}.

As shown in Figure 13, CV of the hydroperoxide complex **32** displayed an irreversible oxidation peak at −0.18 V *versus* Fc$^+$/Fc in 1,2-difluorobenzene at 273 K. According to the authors, the process was monoelectronic and led to the transient Cu(III)-OOH species. Increase of the scan rate until 2 V s^{-1} did not reveal any reversibility, thus suggesting that a fast chemical reaction followed the oxidation of the hydroperoxide species (EC mechanism, E = electrochemical, C = chemical). Interestingly, the peak potential value was close to that found for the Cu(II)-OH analogous complex **11** (Table 2), but here oxidation was fully irreversible whereas it was reversible for **11** in DFB and acetone. From simulated CVs assuming an EC mechanism, a half-wave potential value equal to −0.21 V *versus* Fc$^+$/Fc was considered to calculate the BDFE(OH) of complex **32** knowing the pKa of the Cu(II)-superoxo/Cu(III)-hydroperoxo couple (pKa = 19). The resulting high BDFE(OH) value (87 kcal mol^{-1}) was indicative of the relatively strong oxidative HAA properties of the superoxide species.

With the same pyridine(dicarboxamide) ligand, alkylperoxide Cu(II) complexes **33** and **34** displayed slightly more positive oxidation potential values than that obtained for complex **32** (see Table 3).[32] However, for the alkylperoxide complexes, reversibility was detected for $v > 0.5$ V s^{-1} at room temperature hence evidencing that the alkyl groups stabilize the Cu(III) redox state. The Cu(III) character of the mono-oxidized species was confirmed by UV–vis and EPR spectroscopic studies by using chemical oxidants.

Mononuclear species

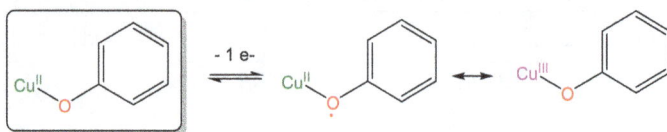

Dinuclear species

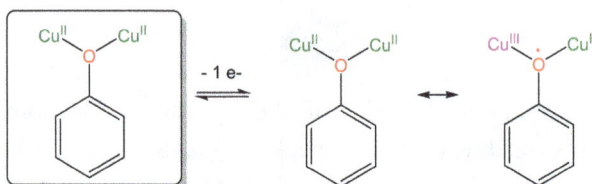

Figure 14. Electron and proton–electron transfer reactions for mono and dinuclear copper phenoxide species. Plain-line rectangles designate the species of interest, here phenolato Cu(II) cores.

V. Electrochemistry of Copper(II)-Phenoxide Species

Phenoxide-based mono- and dicopper(II) precursors have been extensively investigated by electrochemical methods. Indeed, their oxidation affords the generation of phenoxyl Cu(II) radical species (Figure 14), which are of interest for oxidation reactions in synthetic chemistry. Moreover, these species are bio-relevant, since tyrosyl radicals participate as co-factor (tyrosine residue) in biological reactions such as in occurring in galactose oxidase. Many different ligands bearing one or two phenoxide moieties have been reported, and the readers are invited to refer to several articles and reviews that gather spectroscopic and electrochemical properties.[33] In the following, a restricted list of complexes has been considered in the view of their remarkable redox behavior.

For instance, Tolman and co-workers investigated the redox properties of a series of three monocopper(II) complexes bearing a phenoxide moiety attached to a macrocyclic TACN ligand (TACN = triazacyclononane) (complexes **35–38**, Figure 15).[34] These biomimetic complexes were studied at room temperature in dichloromethane. As shown in Table 4, their monoelectronic oxidation occurred at potential values at ca. 0.20 V *versus*

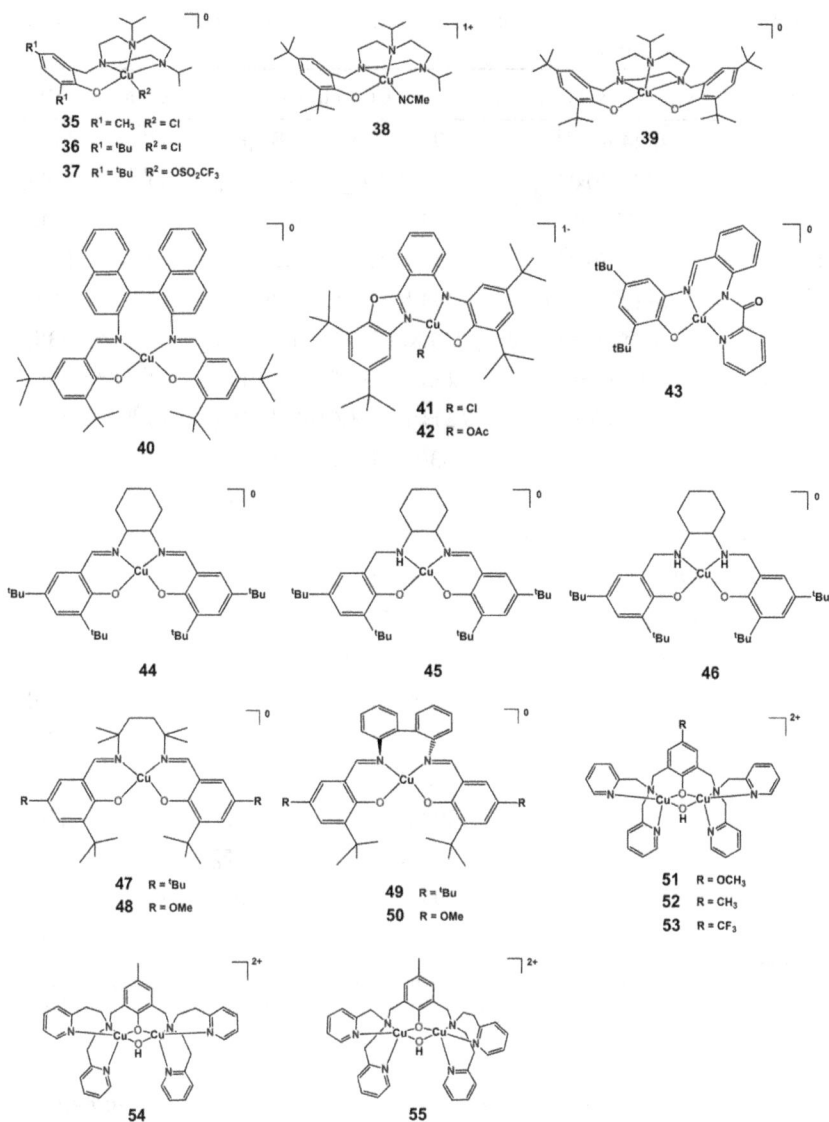

Figure 15. Schematic representation of phenoxo complexes **35–55**.

Fc$^+$/Fc, which were dependent on both the substituents on the phenoxo unit (120 mV increase from Me to iPr) and the exogenous ligand bound to the Cu(II) (chloride or triflate anions). The Cu(II)-phenoxide radicals of complexes **36** and **37** were also prepared by exhaustive electrolysis and

Table 4. Electrochemical data for phenoxo (O^{Ph}) copper(II) complexes **35–55** (red: reduction; ox: oxidation).

Complex	$E_{1/2}$ / V vs. Fc^+/Fc	Conditions	Adduct	Ref.
35	0.24 (ox)[a,b]	293 K, CH_2Cl_2/NBu_4PF_6	O^{Ph}	[34a]
36	0.12 (ox)[b]	293 K, CH_2Cl_2/NBu_4PF_6	O^{Ph}	[34a]
37	0.25 (ox)[b]	293 K, CH_2Cl_2/NBu_4PF_6	O^{Ph}	[34a]
38	0.32 (ox)[b]	293 K, CH_2Cl_2/NBu_4PF_6	O^{Ph}	[34b]
39	0.03 (ox1)[b]; 0.31 (ox2)[b]	293 K, CH_2Cl_2/NBu_4PF_6	O^{Ph}	[34b]
40	0.41 (ox1)[c]; 0.58 (ox2)[c]	233 K, CH_3CN/NBu_4ClO_4	O^{Ph}	[35]
41	0.55 (ox)	233 K, CH_2Cl_2/NBu_4PF_6	O^{Ph}	[36]
42	0.28 (ox)	233 K, CH_2Cl_2/NBu_4PF_6	O^{Ph}	[36]
43	0.46 (ox)	293 K, CH_2Cl_2/NBu_4ClO_4	O^{Ph}	[37]
44	0.45 (ox1); 0.65 (ox2)	293 K, CH_2Cl_2/NBu_4ClO_4	O^{Ph}	[38]
45	0.15 (ox1); 0.54 (ox2)	293 K, CH_2Cl_2/NBu_4ClO_4	O^{Ph}	[38a]
46	0.08 (ox1); 0.21 (ox2)	293 K, CH_2Cl_2/NBu_4ClO_4	O^{Ph}	[38]
47	0.44 (ox1); 0.76 (ox2)	293 K, CH_2Cl_2/NBu_4ClO_4	O^{Ph}	[39]
48	0.22 (ox1); 0.50 (ox2)	293 K, CH_2Cl_2/NBu_4ClO_4	O^{Ph}	[39]
49	0.56 (ox1); 0.75 (ox2)	293 K, CH_2Cl_2/NBu_4ClO_4	O^{Ph}	[39]
50	0.34 (ox1); 0.53 (ox2)	293 K, CH_2Cl_2/NBu_4ClO_4	O^{Ph}	[39]
51	0.51 (ox)	293 K, CH_3CN/NBu_4PF_6	O^{Ph}	[33d,40]
52	0.71 (ox)[a]	293 K, CH_3CN/NBu_4PF_6	O^{Ph}	[40]
53	1.20 (ox)[a]	293 K, CH_3CN/NBu_4PF_6	O^{Ph}	[40]
54	1.04 (ox)[a]	293 K, CH_3CN/NBu_4PF_6	O^{Ph}	[40]
55	0.90 (ox)[a]	293 K, CH_3CN/NBu_4PF_6	O^{Ph}	[40]

[a]Irreversible peak; [b]Recalculated versus Fc^+/Fc by taking $E^0(Fc^+/Fc) = 0.47$ V versus SCE in this solvent; [c]Recalculated versus Fc^+/Fc by taking $E^0(Fc^+/Fc) = 0.40$ V versus SCE in this solvent.

characterized by UV–vis and EPR spectroscopies.[34a] The resulting spectra exhibited a band around 410 nm, which was ascribed to a π-π^* transition of the phenoxyl radicals by analogy with analogous studies with free radicals. In a further paper,[34b] the same research group synthesized a bis-phenoxide complex of the same family of ligand (TACN) (complex **39**). CV studies displayed two reversible monoelectronic waves at 0.03 and

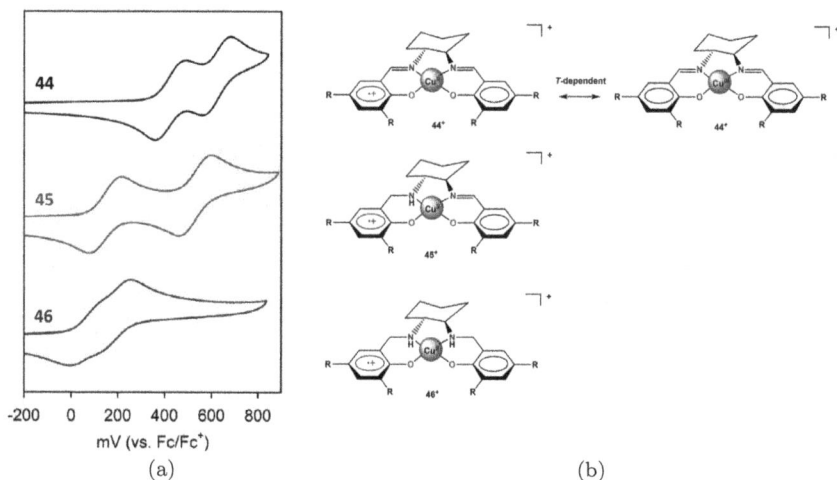

Figure 16. (a) Room-temperature CVs at 0.1 V s^{-1} of complex **44** (black), **45** (red), and **46** (blue) in CH_2Cl_2/NBu_4ClO_4 (R = ^{t}Bu). (b) Schematic representation of the mono-oxidized species **44$^+$**, **45$^+$**, and **46$^+$**. Adapted with permission from Refs. [38a] and [38b]. Copyright (2018) Elsevier. Copyright (2008) American Chemical Society.

0.31 V *versus* Fc$^+$/Fc, which were ascribed to the successive oxidation of each phenoxo moieties by comparison with an analogous bis-phenoxide Zn(II) complex and the phenoxo complex **36**. Chemical generation of the mono-phenoxyl species was achieved by reaction of complex **38** with a Ce(IV) salt at 233 K.

Another interesting study of Cu(II)-phenoxide oxidation was also performed by the group of Stack in 2003 by using salen ligand (N$_2$O$_2$ Schiff bases).[38, 41] Two complexes differing by the nature of the bridging moiety (complexes **44** and **46**, Figure 15) were analyzed by electrochemical and spectroscopic methods. As shown in Table 1, a substantial decrease of the oxidation potential by nearly 400 mV for both oxidation events was observed when the imine groups were substituted by amines (Figure 16). This result was consistent with the stronger basic properties of the amine. However, complex **46** was shown to be unexpectedly better than complex **44** for the catalytic oxidation of benzylic alcohol in presence of a chemical oxidant.[38] More recently, Storr and co-workers

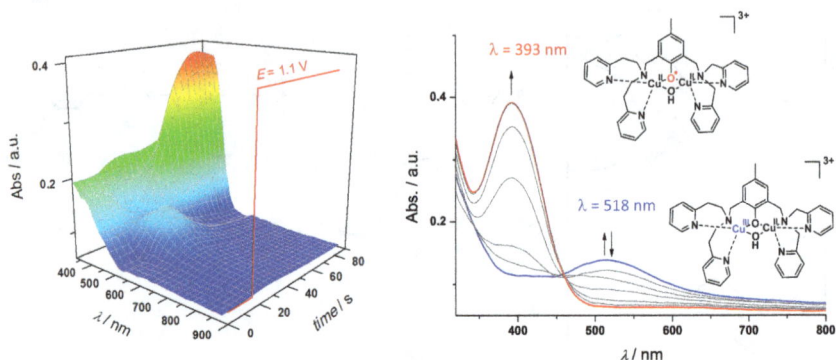

Figure 17. UV — vis spectroelectrochemical monitoring of the oxidation of complex **55** in CH_3CN/NBu_4PF_6. Adapted with permission from Ref. [40]. Copyright (2017) American Chemical Society.

completed these investigations by determining the redox and catalytic properties of complex **45** bearing one imino group and one amino moiety (Figure 15).[38a] They found that mono-oxidation occurred on the aminophenoxide side, leading to the formation of Cu(II)-phenoxyl species. However, benzylic alcohol oxidation was found to be slower than for complex **46**.

Very recently, bis-phenoxide salen monoCu(II) complexes **47–50** (Figure 15) comprising distorted bridging ligands and various substituents on the phenoxide moieties were studied by Thomas and co-workers[39] in the continuation of Stack's complex **40**.[35] As shown in Table 4, a substantial variation of the oxidation potential value for the generation of mono-phenoxyl radical species was observed for both oxidation processes, particularly by changing the donor properties of the substituting groups (OMe vs. tBu) on the phenoxide moieties. The generated mono-Cu(II) mono-phenoxyl radicals, which were all assigned to class II mixed-valent compounds, displayed remarkable catalytic properties for the aerobic oxidation of unactivated primary alcohols such as 2-phenylethanol.

Dicopper(II) complexes **51–55** comprising hydroxo and phenolato bridging moieties as well as bis(alkylpyridyl)amine ligands were thoroughly studied by electrochemistry, spectroelectrochemistry, and DFT calculations.[40] Whereas complex **51** bearing a methoxy-phenolate ligand displayed a reversible monoelectronic oxidation at 0.51 V *versus* Fc^+/Fc, the other Cu(II) complexes **52–55** exhibited an irreversible oxidation peak at potential values higher than 0.71 V. A remarkable result was obtained with the unsymmetrical complex **55** since a transient mixed-valent Cu(II)–Cu(III) phenoxide species displaying a PhO$^-$ → Cu(III) charge transfer band at 518 nm was detected by performing room-temperature *in situ* and time-resolved spectroelectrochemical experiments (Figure 17). This species evolved rapidly ($t_{1/2}$ = 14 s at 20°C) toward a dicopper (II,II) phenoxyl radical complex. The specific redox behavior of this complex was ascribed to the unsymmetrical design of the ligand combined with the chain length separating the amine from the pyridyl groups on one side of the molecule. According to the authors, ethyl spacers (vs. methyl, see complex **52**, Figure 15) were sufficiently flexible to accommodate for the geometrical arrangements required to stabilize the Cu(III) species.

VI. Electrochemistry of Nitroso and Sulfide Analogues of Copper–Oxygen Complexes

Very recent works have shown that nitroso and sulfide Cu(II) adducts could also be of interest for oxidation reaction, such as HAA, by analogy with their copper–oxygen counterparts. Figure 18 displays the different electron-transfer reactions that can be expected from such species. On one hand, mono or dicopper(II) nitrisoarene complexes may undergo monoelectronic reduction since they are isoelectronic analogues of copper superoxide adducts, the electron input taking place either on the metal or on the nitrosoarene ligand. On the other hand, copper(II)-sulfide are analogues of Cu(II)-hydroxo adducts. They can be thus employed as precursors for the generation of high-valent Cu(III) species as for the copper(II)-hydroxo case. In consequence, their redox behavior is expected to be similar to that of Cu(II)-OH species. At last, *side-on* dicopper(II) bisulfides are analogues of dicopper(II) peroxides, hence expecting that their reduction would lead to reactive mixed-valent Cu(II)–Cu(III) bis(μ-S) adducts (Figure 18).

The recent investigations performed by Ottenwaelder and co-workers on copper-nitrisoarene complexes have shed light on the redox behavior of this family of complexes depending on the electronic properties of the nitrosobenzene ligand (complexes **56–59**, Figure 19).[42] These adducts

Copper nitrosoarene species

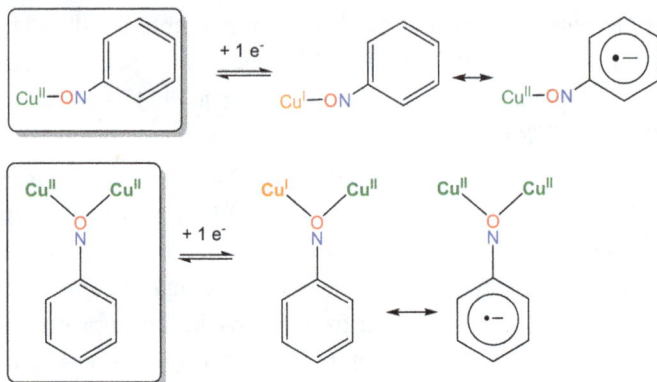

Copper sulfide and disulfide species

R=H, Ph

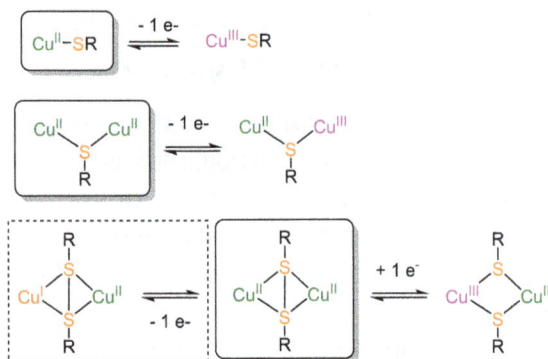

Figure 18. Electron transfer reactions for mono and dinuclear nitrosoarene and (bi) sulfido copper species. Plain-line rectangles designate the species of interest, here nitrosoarene, sufido, and bisulfido Cu(II) cores. The dotted-line rectangle indicates a reaction that has not yet been evidenced by electrochemical methods.

were prepared by reaction of copper(I) complexes of N,N,N',N'-tetramethypropylenediamine ligand (TMPD) with a series of *para*-substituted nitrosobenzene derivatives. Through adequate tuning of the substituting groups on the arylnitroso moiety, various Cu-NO binding modes were obtained and characterized at the solid state and in solution by spectroscopic and electrochemical methods. Voltammetric studies of

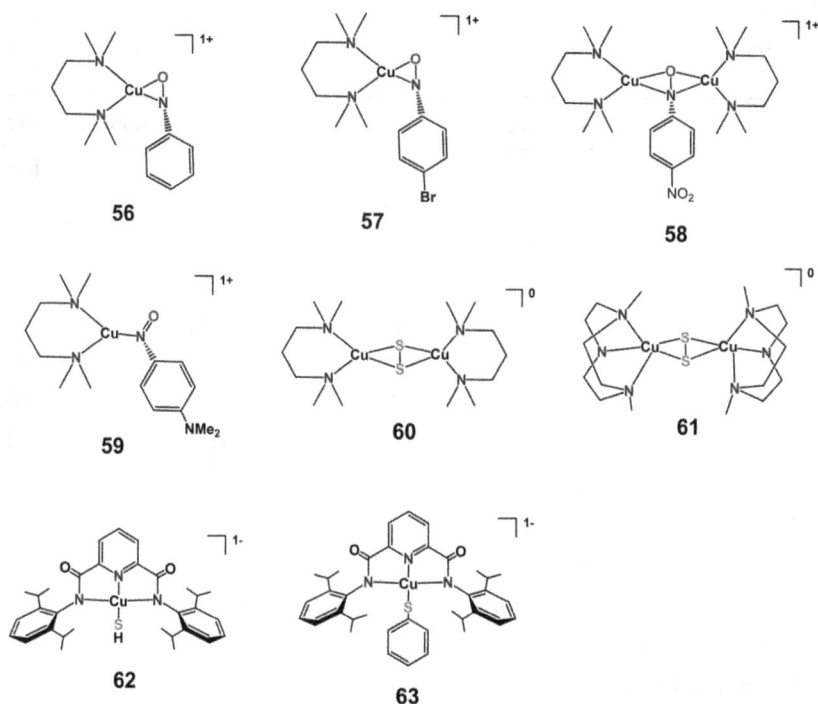

Figure 19. Schematic representation of nitroso, disulfido, hydrosulfido, and thiophenolato complexes **56–63**.

both the redox-active nitrosoarene ligands and their copper complexes were carried out in dichloromethane. The ligands exhibited two successive quasi-reversible systems when scanning negatively, consistent with an ECE mechanism comprising two monoelectronic electron transfers. In strong contrast, all copper complexes displayed an irreversible reduction peak at potential values varying between −0.72 V and −1.05 V *versus* $Fc^+/$ Fc according to the donating/withdrawing properties of the substituent (Table 5). These potential values were roughly 500 mV more positive than those found for the nitrosoarene ligands. Exhaustive electrolysis indicated that the reduction processes were monoelectronic and led to fast decomplexation, which released either the TMPD or nitrosoarene moieties. On the other hand, electrochemical oxidation of the copper–nitrosoarene complexes yielded a quasi-reversible system at a potential value varying

Table 5. Electrochemical data for nitrosoarene (^{PhNO}Cu), *side-on* disuldido (^{S}S), hydrosulfido (HS), and thiophenolato (PhS) complexes **56–63** (red: reduction; ox: oxidation).

Complex	$E_{1/2}$ / V $vs.$ Fc$^+$/Fc	Conditions	Adduct	Ref.
56	−0.92 (red)^a; 0.28 (ox)^a	293 K, CH$_2$Cl$_2$/NBu$_4$OTf	^{PhNO}Cu	[42]
57	−0.82 (red)^a; 0.08 (ox)	293 K, CH$_2$Cl$_2$/NBu$_4$OTf	^{PhNO}Cu	[42]
58	−0.72 (red)^a; 0.11 (ox)	293 K, CH$_2$Cl$_2$/NBu$_4$OTf	^{PhNO}Cu	[42]
59	−1.05 (red)^a; −0.36 (ox)	293 K, CH$_2$Cl$_2$/NBu$_4$OTf	^{PhNO}Cu	[42]
60	−0.30 (red)^a	293 K, CH$_2$Cl$_2$/NBu$_4$PF$_6$	^{S}S	[43]
61	−0.77 (red)^a	293 K, CH$_2$Cl$_2$/NBu$_4$PF$_6$	^{S}S	[43]
62	−0.21 (ox)	293 K, THF/NBu$_4$PF$_6$	HS	[44]
63	−0.25 (ox)	293 K, THF/NBu$_4$PF$_6$	PhS	[44]

^{a}Irreversible peak.

between −0.36 and 0.28 V *versus* Fc$^+$/Fc, depending on the nature of substituting group. These complexes were shown to be capable to perform H-atom abstraction on C–H bonds of dihydroanthracene (BDFE = 76 kcal mol^{-1}) but at low reaction rates.

Another electrochemical study of interest was carried out by Tolman and co-workers in 2008 on *side-on* disulfido dicopper complexes,[43] which are structural models of *side-on* peroxide copper adducts. Voltammetry at room temperature in dichloromethane displayed an irreversible reduction peak at −0.30 V and −0.77 V *versus* Fc$^+$/Fc for complexes **60** and **61**, respectively (Figure 19 and Table 5). The difference of potential was ascribed to the donor ability of the neutral ligand (peralkylated bi- or tri-amine). In contrast to complex **61**, complex **60** was shown to be a two-electron oxidant for substituted phenolates. More recently, the same group conducted researches on sulfur-containing analogues of monocopper-hydroxo species.[44] Two Cu(II) complexes bearing a macrocyclic ligand and a hydrosulfido and thiophenolato anion were synthesized (Figure 19, complexes **62** and **63**, respectively). Room-temperature voltammetric studies of these complexes in THF exhibited a quasi-reversible 1e$^-$ system in oxidation at −0.21 V and −0.25 V *versus* Fc$^+$/Fc (Table 5). The slightly more negative $E_{1/2}$ value for the complex **63** *versus* **62** was consistent with the better donor properties of the phenyl moiety compared to the hydrogen atom. Analogously, complex **62** showed a significantly lower oxidation potential value (−50 mV) than that obtained for its hydroxo counterpart (complex **11**), as a result of the higher electronegativity of S *versus* O. Reactivity of complexes **62** and **63** with TEMPOH was much slower than for the OH analogue (complex **11**) in agreement with the electrochemical data and the lower basicity of the SH⁻ and SPh⁻ ligands relative the hydroxide.

VII. Summary

In conclusion, this chapter has reviewed the electrochemical and spectro-electrochemical studies of relevant copper species, including unstable copper oxygen adducts and their precursors. While the reported examples remain still scarce, one can observe that the number of electrochemical studies of these species has dramatically increased for the last five years. One main reason is that electrochemical data (thermodynamics) can be directly correlated to HAA reactivity, hence allowing the calculation of BDFEs through the well-known Bordwell analysis knowing pKa values. Moreover, under certain circumstances, kinetics of electron transfer have been directly quantified from electrochemical measurements (CV or EIS), giving the possibility to determine inner and outer reorganizational energies with the help of heterogeneous electron transfer theories. Another source of progress has been achieved with *in situ* UV–visible cryo-spectroelectrochemical measurements that have afforded direct characterization of transient species generated at the electrode surface at low temperature[21,28,45] hence avoiding exhaustive electrolysis and uncertainty about the real nature of the generated transient species.

From all data given in this chapter, it seems difficult to provide definitive conclusions and/or to predict any redox behavior for any or other specific family of copper–oxygen adducts. The principal reason is that subtle variation of the geometric or electronic properties of the ligand itself can induce significant modification of the redox properties, such as the half-wave potential value, or even move the location of the electron transfer. This is best exemplified by the different UV–vis spectroelectrochemical responses that were obtained upon oxidation of the phenoxo-based complexes **52** and **55**, which only differ by the spacer chain length on one side of the dinucleating ligand.

In the future, the employment of electrochemistry and spectroelectro-chemistry for investigating the reactivity of copper–oxygen adducts may become more systematic given the clear benefits provided by such meas-urements. In complement, the development of new coupled *in situ* spectroelectrochemical techniques adapted to this chemistry, such as EPR and resonance Raman, may be of great help to better rationalize their structure-reactivity relationships. Furthermore, the remarkable oxidative properties of these species have not yet been fully exploited for the homo-geneous and heterogeneous electrocatalytic oxidation of organic sub-strates. Efforts in these directions should provide substantial progress toward a better utilization of their remarkable properties.

VIII. References

1. (a) Solomon, E. I.; Heppner, D. E.; Johnston, E. M.; Ginsbach, J. W.; Cirera, J.; Qayyum, M.; Kieber-Emmons, M. T.; Kjaergaard, C. H.; Hadt, R. G.; Tian, L. *Chem. Rev.* **2014**, *114*, 3659–3853. (b) Quist, D. A.; Diaz, D. E.; Liu, J. J.; Karlin, K. D. *J. Biol. Inorg. Chem.* **2017**, *22*, 253–288. (c) Lee, J. Y.; Karlin, K. D. *Curr. Opin. Chem. Biol.* **2015**, *25*, 184–193.

2. Meier, K. K.; Jones, S. M.; Kaper, T.; Hansson, H.; Koetsier, M. J.; Karkehabadi, S.; Solomon, E. I.; Sandgren, M.; Kelemen, B. *Chem. Rev.* **2018**, *118*, 2593–2635.

3. Cowley, R. E.; Tian, L.; Solomon, E. I. *Proc. Natl. Acad. Sci. U.S.A.* **2016**, *113*, 12035–12040.

4. Solomon, E. I. *Inorg. Chem.* **2016**, *55*, 6364–6375.

5. Elwell, C. E.; Gagnon, N. L.; Neisen, B. D.; Dhar, D.; Spaeth, A. D.; Yee, G. M.; Tolman, W. B. *Chem. Rev.* **2017**, *117*, 2059–2107.

6. (a) Lyu, Z.; Zhou, Y.; Dai, W.; Cui, X.; Lai, M.; Wang, L.; Huo, F.; Huang, W.; Hu, Z.; Chen, W. *Chem. Soc. Rev.* **2017**. (b) Kwabi, D. G.; Bryantsev, V. S.; Batcho, T. P.; Itkis, D. M.; Thompson, C. V.; Shao-Horn, Y. *Angew. Chem. Int. Ed.* **2016**, *55*, 3129–3134. (c) Bruce, P. G.; Freunberger, S. A.; Hardwick, L. J.; Tarascon, J. M. *Nat. Mater.* **2011**, *11*, 19–29.

7. (a) Poulos, T. L. *Chem. Rev.* **2014**, *114*, 3919–3962. (b) Sheng, Y.; Abreu, I. A.; Cabelli, D. E.; Maroney, M. J.; Miller, A. F.; Teixeira, M.; Valentine, J. S. *Chem. Rev.* **2014**, *114*, 3854–3918.

8. (a) Dhar, D.; Tolman, W. B. *J. Am. Chem. Soc.* **2015**, *137*, 1322–1329. (b) Warren, J. J.; Tronic, T. A.; Mayer, J. M. *Chem. Rev.* **2010**, *110*, 6961–7001.

9. (a) Mandal, M.; Elwell, C. E.; Bouchey, C. J.; Zerk, T. J.; Tolman, W. B.; Cramer, C. J. *J. Am. Chem. Soc.* **2019**, *141*, 17236–17244. (b) Bailey, W. D.; Dhar, D.; Cramblitt,

A. C.; Tolman, W. B. *J. Am. Chem. Soc.* **2019**, *141*, 5470–5480. (c) Kindermann, N.; Gunes, C. J.; Dechert, S.; Meyer, F. *J. Am. Chem. Soc.* **2017**, *139*, 9831–9834. (d) VanNatta, P. E.; Ramirez, D. A.; Velarde, A. R.; Ali, G.; Kieber-Emmons, M. T. *J. Am. Chem. Soc.* **2020**, *142*, 16292–16312. (e) Ali, G.; VanNatta, P. E.; Ramirez, D. A.; Light, K. M.; Kieber-Emmons, M. T. *J. Am. Chem. Soc.* **2017**, *139*, 18448–18451.

10. Mano, N.; de Poulpiquet, A. *Chem. Rev.* **2018**, *118*, 2392–2468.

11. (a) Thorseth, M. A.; Tornow, C. E.; Tse, E. C. M.; Gewirth, A. A. *Coord. Chem. Rev.* **2013**, *257*, 130–139. (b) Langerman, M.; Hetterscheid, D. G. H. *Angew. Chem. Int. Ed.* **2019**, *58*, 12974–12978. (c) Gentil, S.; Serre, D.; Philouze, C.; Holzinger, M.; Thomas, F.; Le Goff, A. *Angew. Chem. Int. Ed. Engl.* **2016**, *55*, 2517–2520. (d) Thorseth, M. A.; Letko, C. S.; Rauchfuss, T. B.; Gewirth, A. A. *Inorg. Chem.* **2011**, *50*, 6158–6162. (e) Kato, M.; Yagi, I. *e-J. Surf. Sci. Nanotech.* **2020**, *18*, 81–93.

12. (a) Lukács, D.; Szyrwiel, Ł.; Pap, J. *Catalysts* **2019**, *9*, 83. (b) Koepke, S. J.; Light, K. M.; VanNatta, P. E.; Wiley, K. M.; Kieber-Emmons, M. T. *J. Am. Chem. Soc.* **2017**, *139*, 8586–8600. (c) Kafentzi, M. C.; Papadakis, R.; Gennarini, F.; Kochem, A.; Iranzo, O.; Le Mest, Y.; Le Poul, N.; Tron, T.; Faure, B.; Simaan, A. J.; Reglier, M. *Chem. Eur. J.* **2018**, *24*, 5213–5224. (d) Barnett, S. M.; Goldberg, K. I.; Mayer, J. M. *Nat. Chem.* **2012**, *4*, 498–502. (e) Su, X. J.; Gao, M.; Jiao, L.; Liao, R. Z.; Siegbahn, P. E.; Cheng, J. P.; Zhang, M. T. *Angew. Chem. Int. Ed.* **2015**, *54*, 4909–4914. (f) Nestke, S.; Ronge, E.; Siewert, I. *Dalton Trans.* **2018**, *47*, 10737–10741.

13. Tahsini, L.; Kotani, H.; Lee, Y.-M.; Cho, J.; Nam, W.; Karlin, K. D.; Fukuzumi, S. *Chem. Eur. J.* **2012**, *18*, 1084–1093.

14. Mirica, L. M.; Ottenwaelder, X.; Stack, T. D. P. *Chem. Rev.* **2004**, *104*, 1013–1046.

15. (a) Cao, R.; Saracini, C.; Ginsbach, J. W.; Kieber-Emmons, M. T.; Siegler, M. A.; Solomon, E. I.; Fukuzumi, S.; Karlin, K. D. *J. Am. Chem. Soc.* **2016**, *138*, 7055–7066. (b) Dhar, D.; Yee, G. M.; Spaeth, A. D.; Boyce, D. W.; Zhang, H.; Dereli, B. Cramer, C. J.; Tolman, W. B. *J. Am. Chem. Soc.* **2016**, *138*, 356–368.

16. Gagnon, N.; Tolman, W. B. *Acc. Chem. Res.* **2015**, *48*, 2126–2131.

17. Shiota, Y.; Yoshizawa, K. *Inorg. Chem.* **2009**, *48*, 838–845.

18. Karlin, K. D.; Tyeklar, Z.; Farooq, A.; Haka, M. S.; Ghosh, P.; Cruse, R. W.; Gultneh, Y.; Hayes, J. C.; Toscano, P. J.; Zubieta, J. *Inorg. Chem.* **1992**, *31*, 1436–1451.

19. Lopez, I.; Porras-Gutierrez, A. G.; Douziech, B.; Wojcik, L.; Le Mest, Y.; Kodera, M.; Le Poul, N. *Chem. Commun.* **2018**, *54*, 4931–4934.

20. Shearer, J.; Zhang, C. X.; Zakharov, L. N.; Rheingold, A. L.; Karlin, K. D. *J. Am. Chem. Soc.* **2005**, *127*, 5469–5483.

21. Lopez, I.; Cao, R.; Quist, D. A.; Karlin, K. D.; Le Poul, N. *Chem. Eur. J.* **2017**, *23*, 18314–18319.

22. De Leener, G.; Over, D.; Smet, C.; Cornut, D.; Porras-Gutierrez, A. G.; Lopez, I.; Douziech, B.; Le Poul, N.; Topic, F.; Rissanen, K.; Le Mest, Y.; Jabin, I.; Reinaud, O. *Inorg. Chem.* **2017**, *56*, 10971–10983.

23. Halvagar, M. R.; Solntsev, P. V.; Lim, H.; Hedman, B.; Hodgson, K. O.; Solomon, E. I.; Cramer, C. J.; Tolman, W. B. *J. Am. Chem. Soc.* **2014**, *136*, 7269–7272.

24. (a) Donoghue, P. J.; Tehranchi, J.; Cramer, C. J.; Sarangi, R.; Solomon, E. I.; Tolman, W. B. *J. Am. Chem. Soc.* **2011**, *133*, 17602–17605. (b) Krishnan, V. M.; Shopov, D. Y.; Bouchey, C. J.; Bailey, W. D.; Parveen, R.; Vlaisavljevich, B.; Tolman, W. B. *J. Am. Chem. Soc.* **2021**, *143*, 3295–3299.

25. Dhar, D.; Yee, G. M.; Tolman, W. B. *Inorg. Chem.* **2018**, *57*, 9794–9806.

26. Elwell, C. E.; Mandal, M.; Bouchey, C. J.; Que, L.; Cramer, Jr. C. J.; Tolman, W. B. *Inorg. Chem.* **2019**, *58*, 15872–15879.

27. Kochem, A.; Gennarini, F.; Yemloul, M.; Orio, M.; Le Poul, N.; Riviere, E.; Giorgi, M.; Faure, B.; Le Mest, Y.; Reglier, M.; Simaan, A. J. *Chem. Plus. Chem.* **2017**, *82*, 615–624.

28. Thibon-Pourret, A.; Gennarini, F.; David, R.; Isaac, J. A.; Lopez, I.; Gellon, G.; Molton, F.; Wojcik, L.; Philouze, C.; Flot, D.; Le Mest, Y.; Reglier, M.; Le Poul, N.; Jamet, H.; Belle, C. *Inorg. Chem.* **2018**, *57*, 12364–12375.

29. Wu, T.; MacMillan, S. N.; Rajabimoghadam, K.; Siegler, M. A.; Lancaster, K. M.; Garcia-Bosch, I. *J. Am. Chem. Soc.* **2020**, *142*, 12265–12276.

30. Isaac, J. A.; Gennarini, F.; Lopez, I.; Thibon-Pourret, A.; David, R.; Gellon, G.; Gennaro, B.; Philouze, C.; Meyer, F.; Demeshko, S.; Le Mest, Y.; Reglier, M.; Jamet, H.; Le Poul, N.; Belle, C. *Inorg. Chem.* **2016**, *55*, 8263–8266.

31. Isaac, J. A.; Thibon-Pourret, A.; Durand, A.; Philouze, C.; Le Poul, N.; Belle, C. *Chem. Commun.* **2019**, *55*, 12711–12714.

32. Neisen, B. D.; Gagnon, N. L.; Dhar, D.; Spaeth, A. D.; Tolman, W. B. *J. Am. Chem. Soc.* **2017**, *139*, 10220–10223.

33. (a) Thomas, F. *Eur. J. Inorg. Chem.* **2007**, *2007*, 2379–2404. (b) Benisvy, L.; Blake, A. J.; Davies, E. S.; Garner, C. D.; McMaster, J.; Wilson, C.; Collison, D.; McInnes, E. J. L.; Whittaker, G. *Chem. Commun.* **2001**, 1824–1825. (c) Uma Maheswari, P.; Hartl, F.; Quesada, M.; Buda, F.; Lutz, M.; Spek, A. L.; Gamez, P.; Reedijk, J. *Inorg. Chim. Acta* **2011**, *374*, 406–414. (d) Michel, F.; Torelli, S.; Thomas, F.; Duboc, C.; Philouze, C.; Belle, C.; Hamman, S.; Saint-Aman, E.; Pierre, J. L. *Angew. Chem. Int. Ed.* **2005**, *44*, 438–441.

34. (a) Halfen, J. A.; Young, V. G.; Tolman, W. B. *Angew. Chem. Int. Ed.* **1996**, *35*, 1687–1690. (b) Halfen, J. A.; Jazdzewski, B. A.; Mahapatra, S.; Berreau, L. M.; Wilkinson, E. C.; Que, L.; Tolman, W. B. *J. Am. Chem. Soc.* **1997**, *119*, 8217–8227.

35. Wang, Y.; Stack, T. D. P. *J. Am. Chem. Soc.* **1996**, *118*, 13097–13098.

36. Balaghi, S. E.; Safaei, E.; Chiang, L.; Wong, E. W.; Savard, D.; Clarke, R. M.; Storr, T. *Dalton Trans.* **2013**, *42*, 6829–6839.

37. Alaji, Z.; Safaei, E.; Chiang, L.; Clarke, R. M.; Mu, C.; Storr, T. *Eur. J. Inorg. Chem.* **2014**, *2014*, 6066–6074.

38. (a) Chiang, L.; Wasinger, E. C.; Shimazaki, Y.; Young, V.; Storr, T.; Stack, T. D. P. *Inorg. Chim. Acta* **2018**, *481*, 151–158. (b) Storr, T.; Verma, P.; Pratt, R. C.; Wasinger, E. C.; Shimazaki, Y.; Stack, T. D. *J. Am. Chem. Soc.* **2008**, *130*, 15448–15459.

39. Kunert, R.; Philouze, C.; Berthiol, F.; Jarjayes, O.; Storr, T.; Thomas, F. *Dalton Trans.* **2020**, *49*, 12990–13002.

40. Gennarini, F.; David, R.; Lopez, I.; Le Mest, Y.; Reglier, M.; Belle, C.; Thibon-Pourret, A.; Jamet, H.; Le Poul, N. *Inorg. Chem.* **2017**, *56*, 7707–7719.

41. Pratt, R. C.; Stack, T. D. *J. Am. Chem. Soc.* **2003**, *125*, 8716–8717.

42. Askari, M. S.; Effaty, F.; Gennarini, F.; Orio, M.; Le Poul, N.; Ottenwaelder, X. *Inorg. Chem.* **2020**, *59*, 8678–8689.

43. Bar-Nahum, I.; York, J. T.; Young, V. G.; Tolman, Jr. W. B. *Angew. Chem. Int. Ed.* **2008**, *47*, 533–536.

44. Wu, W.; De Hont, J. T.; Parveen, R.; Vlaisavljevich, B.; Tolman, W. B. *Inorg. Chem.* **2021**, *60*, 5217–5223.

45. López, I.; Le Poul, N. *Coord. Chem. Rev.* **2021**, *436*, 213823.

6 Inorganic Models of Lytic Polysaccharide Monooxygenases

Ivan Castillo*

*Instituto de Química, Universidad Nacional Autónoma de México,
Circuito Exterior, CU, 04510, Ciudad de México, México
joseivan@unam.mx

I. Introduction

Global sources of energy and chemicals have depended heavily on petroleum-based platforms, which have been affordable thus far, at least from an economic standpoint. It is evident, however, that sustainable practices need to be prioritized to avoid serious environmental consequences, beyond the ones we experience nowadays.[1] To this end, efficient use of renewable carbon feedstocks is paramount for the sustainable production of fuels and chemical supplies.[2,3]

The aim of this chapter is to present a brief background of the discovery of copper-dependent lytic polysaccharide monooxygenases (LPMOs),[4-6] within the context of inorganic and coordination chemistry. Their discovery and characterization have led to the design of synthetic systems that possess some of the properties of the enzymatic active sites. Copper complexes thus developed will be discussed in terms of their capability to act as LPMO mimics, and to provide insight on the mechanism

187

that allows the oxidative degradation of recalcitrant polysaccharides and model substrates, starting with representative examples of copper-based hydrolase mimics.

II. Lytic Polysaccharide Monooxygenases

The family of copper-dependent LPMOs was discovered among a consortium of enzymes that act collectively to degrade recalcitrant polysaccharides such as cellulose and chitin. Fungi and bacteria use such LPMO enzymes to oxidatively cleave polysaccharide chains on their natural solid substrates, in contrast to the more commonly described hydrolysis carried out by numerous hydrolase enzymes.[7,8]

The active site of LPMOs consists of a type 2 copper center exposed on the flat face of the enzymes that allows the interaction with its crystalline polymeric substrates.[9] The coordination environment is defined by three nitrogen donors, two of them from imidazole histidines and the third one from the amino terminus that is part of one of the histidine residues. This chelating motif has also been observed in copper-dependent particulate methane monooxygenase and is referred to as the "histidine brace," see Figure 1.

The electronic environment dictated by the T-shaped N_3 donor set is one of the key elements necessary to emulate the physicochemical properties of the LPMO active sites. These include among others the coordination geometry for both Cu^I and Cu^{II}, the redox potential, and ultimately the reactivity toward O_2 and H_2O_2 as oxidants to generate a yet to be identified highly reactive copper–oxygen species. Aside from the T-shape

R = H, Me; X = H, OH

Figure 1. Schematic representation of the reduced (Cu^I) active site of LPMOs.

environment that defines the equatorial coordination plane around copper centers, and the bidentate histidine brace motif, another noteworthy aspect of the LPMOs is the proximity of a phenylalanine or tyrosine residue to one of the axial positions of the copper ions.[4,10,11] Additionally, in fungal enzymes, an unusual methylation of the amino-terminal histidine imidazole occurs at the N-atom.

Polysaccharide hydrolysis is hampered by their tightly packed crystallinity, and their recalcitrant nature is further reflected in the conditions required to achieve oxidative degradation. In this regard, a potent oxidant must be generated at the active site of LPMOs to activate the very strong C–H bond at C1 or C4 of the polymer chains in the first step (Figure 2),[12] which has an estimated strength of 95 kcal mol^{-1}.[13] This leads to aldonic ketones and aldonolactones as products, with the latter being easily hydrolyzed to the corresponding aldonic acids in aqueous solution. These products constitute the oxidized end of the broken polymer chain, while the remaining polymer should remain intact, as shown in Figure 2.

It has been speculated whether the natural oxidant employed by LPMOs consists of atmospheric O_2, or H_2O_2 generated by the microorganisms.[14] In the latter case, it might be worth considering renaming the enzymes Lytic Polysaccharide Peroxygenases (LPPO). Furthermore, the

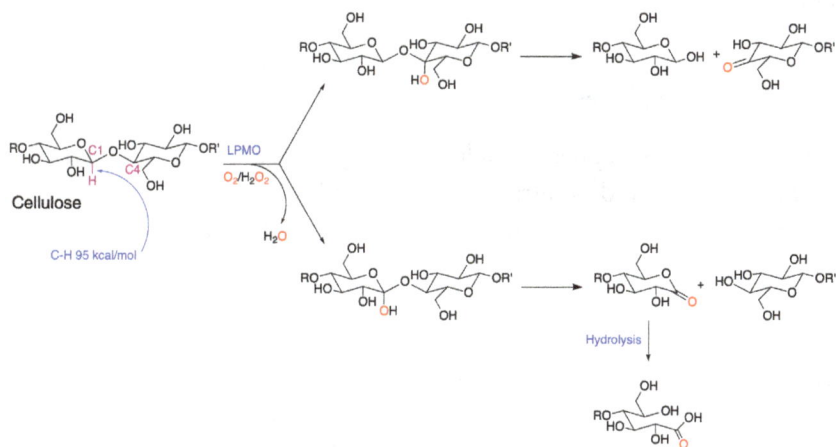

Figure 2. Oxidative mechanism of cellulose cleavage by LPMOs: top, aldonic ketone formation after activation at C4; bottom, aldonolactone formation after activation at C1, followed by hydrolysis. R and R′ represent the polymer chains.

nature of the reactive copper–oxygen species responsible for the initial activation of the C–H bond at C1 or C4 of the monomer that approaches the active site of the enzymes remains to be established.[15] Both questions may be answered by combining the insight gained from interrogating the enzymes themselves, with the information that simple inorganic models have provided in other copper–oxygen systems. A better understanding of the mechanistic details of recalcitrant polysaccharide degradation should lead to better catalysts for industrial applications to develop sustainable, second-generation biofuels and chemicals from the abundant sources that cellulose, chitin, and lignocellulose represent.

III. Copper Complexes as Hydrolase Mimics

A. Polysaccharide Hydrolysis

In contrast to the oxidative cleavage represented in Figure 2, initial polysaccharide degradation studies with copper complexes as catalysts involved a glycoside hydrolysis. In this type of reactivity, carried out enzymatically by hydrolases, attack of a molecule of water at either C1 or C4 should result in the same products. Thus, no oxidation takes place at either end of the broken polymer chains, as exemplified schematically for chitin in Figure 3.

B. Copper Complexes

Studies regarding copper complexes as hydrolase mimics involve the commonly employed p-nitrophenylpyranosides model substrates, with cleavage of the p-nitrophenolate moiety conveniently monitored by

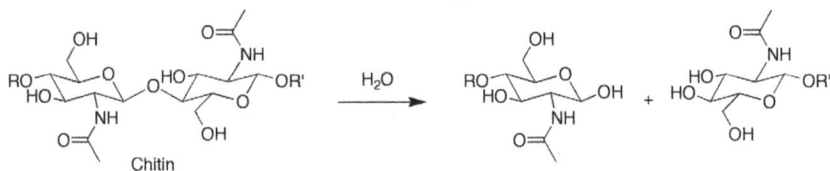

Figure 3.   Hydrolytic mechanism of chitin cleavage; R and R′ represent the polymer chains.

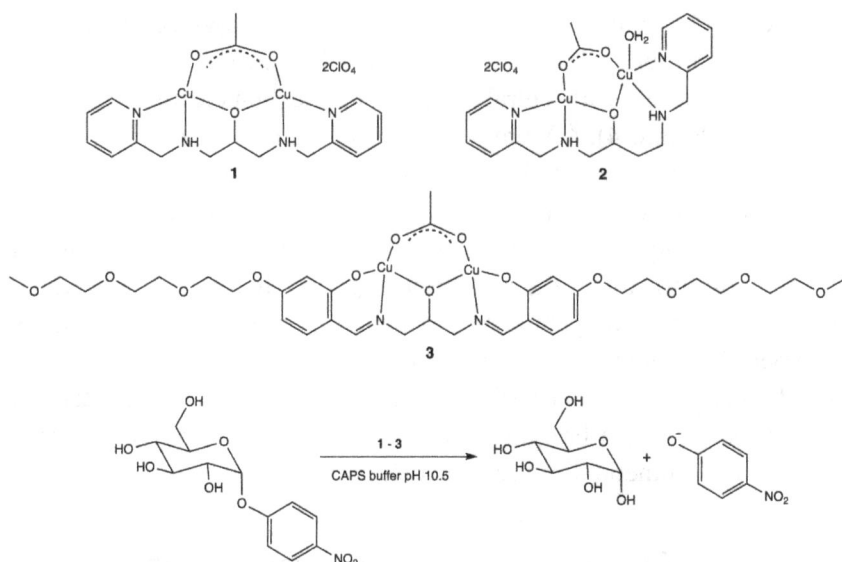

Figure 4. Top: binuclear Cu^{II} complexes **1–3**, bottom: catalytic p-nitrophenyl-α-D-galactopyranoside hydrolysis in aqueous 3-(cyclohexylamino)-1-propanesulfonic acid (CAPS) buffer.

ultraviolet–visible spectroscopy (UV–vis) absorption spectroscopy at 410 nm (Figure 4). This strategy established that binuclear Cu^{II} complexes with N$_2$O donor sets are well suited for pyranoside hydrolysis at pH 10.5, with a 11,000-fold increase relative to the background reaction at such pH value.[16]

The binucleating scaffolds employed in these hydrolase mimics have been exploited for the development of "microgel" particles that incorporate catalytic dicopper sites.[17] Although the microgels display improved hydrolase efficiency, likely due to the use of a carbohydrate template for substrate recognition, the key component appears to be the bimetallic nature of the copper centers. This is reflected in related dicopper systems that display similar reactivity, with only minor modifications from N$_2$O to a N$_2$O$_2$ donor set relative to the original reports.[18] Based on the evidence provided by these dicopper systems, binuclearity appears as a requisite for hydrolysis to occur, as opposed to the oxidative cleavage observed for LPMOs and their models described below.

Chemoselectivity for the hydrolysis of α- and β-glycopyranosides featuring p-nitrophenolate and o-Cl-p-nitrophenolate as leaving groups has been achieved with chiral versions of the binucleating ligand in complex **2** (Figure 4).[19] Moreover, a closer analogy to polysaccharide cleavage was reported with this chiral system, resulting in catalytic hydrolysis of the disaccharides maltose, cellobiose, and lactose at 60°C and pH 8; notably, cellobiose was the most resilient of the disaccharides toward hydrolysis under the conditions reported. In a related report, the cyclic *pseudo*-peptide petallamide, isolated from blue algae, produces dicopper(II) complexes with broad hydrolytic capacity that extends to α- and β-glycosides.[20] Interestingly, one of the structurally characterized complexes features a bridging hydroxo ligand, which may be another requisite for synthetic dicopper glycosidase mimics.

IV. Copper Complexes as LPMO Mimics

Reports concerning copper complexes specifically designed to emulate some properties of the active site of LPMOs, including reactivity, have slowly started to accumulate since the discovery of the enzymes. Such LPMO model systems will be presented chronologically in the next section. A related example of a copper-based system with hydrogen peroxide as oxidant for the degradation of biomass feedstocks in alkaline media exists.[21] Nonetheless, the Cu^{II}/2,2'-bipyridine mixture employed was not characterized, so that the nature of the complex formed in solution remains to be established. It is hard to envision, however, that the species formed may resemble the active site of LPMOs due to the bidentate nature of bipyridine, and the lack of an aliphatic amine.

A. Copper(II)/Bis(Benzimidazolyl)Amine System

Structural analogs of the active site of LPMO enzymes were reported as early as 2014, with the tridentate bis(benzimidazolyl)amine ligand **2BB** in Figure 5 providing the T-shaped N_3 set that includes the amine as central donor.[22] An important feature of the ligands employed for LPMO mimics is the size of the chelate defined by the cis-N_2 moieties: the two

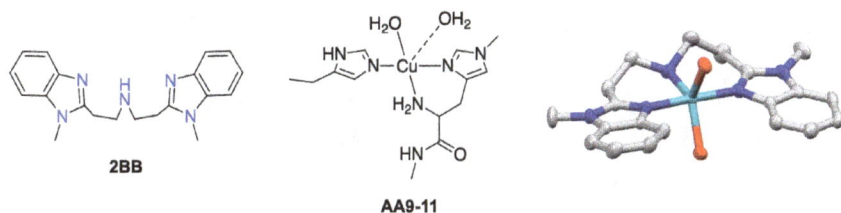

Figure 5. Left: **2BB**; center: schematic representation of the T-shape active site of AA9-11 LPMOs; right: Mercury diagram of [(**2BB**)Cu(H$_2$O)$_2$](OTf)$_2$ at the 50% probability level; H atoms and triflate counterions omitted for clarity. Color code: C, gray; N, blue; O, red; Cu, turquoise.

methylene linkers that connect the central amine to the 2-benzimidazolyl fragments in **2BB** give rise to six-member rings when coordinated to the copper ion. Their relevance lies in the control of the redox potential that the chelate size offers, where five-member rings favor low (more negative) and six-member rings favor high (more positive) potentials.[23] Although electrochemistry was not explored in this initial report, the solid-state structures of the complexes obtained with Cu(BF$_4$)$_2$(H$_2$O)$_6$, CuCl$_2$(H$_2$O)$_2$, and Cu(OTf)$_2$(H$_2$O)$_6$ (OTf$^-$ = CF$_3$SO$_3^-$) as Cu^{II} sources consist of bimetallic [(**2BB**)Cu(μ-F)$_2$](BF$_4$)$_2$, as well as monometallic [(**2BB**)CuCl$_2$] and [(**2BB**)Cu(H$_2$O)$_2$](OTf)$_2$.

Bond lengths in the monometallic complexes [(**2BB**)CuCl$_2$] and [(**2BB**)Cu(H$_2$O)$_2$](OTf)$_2$, in particular, are comparable to those determined for the oxidized (Cu^{II}) forms of the enzymes.[5,24] The Cu–N distances to the benzimidazole donors are almost constant, with an average value of 1.99 Å, while the average Cu–N bond length to the central amine is 2.11 Å; the corresponding Cu–N average distance to the imidazole N-donors from histidine residues is 2.0 Å, and the distance to the primary amines is 2.3 Å. These similarities extend to the T-shape provided by the N$_3$ donor set of **2BB**, and the N–Cu–N angles involved: the *trans* angle involving the cupric center and the benzimidazole N-atoms in the complexes has an average of 170°, while the *cis* angle between the central amine and the benzimidazole donors has a value of 92°. In the enzymes, the values correspond to 165° and 97°, respectively. Despite these similarities, [(**2BB**)CuCl$_2$] and [(**2BB**)Cu(H$_2$O)$_2$](OTf)$_2$ feature trigonal bipyramidal (TBP) geometries defined by the external Cl^- and H_2O ligands, as opposed to the mostly square pyramidal (SP) geometry of the active site of LPMOs, although in some cases intermediate structures have been observed. The TBP geometries persist in acetonitrile solution, as evidenced by low-temperature electron paramagnetic resonance (EPR) spectroscopy at X-band frequency. The parameters determined for both complexes are virtually identical, with $g_\perp = 2.157$ and $g_\parallel = 2.021$, consistent with a dz^2 ground state; the $A_\parallel$ value could not be determined due to poor spectral resolution.

This work was extended to the cuprous complex [(2BB)Cu]OTf, which is monomeric in solution based on ^1H nuclear magnetic resonance spectroscopy (^1H NMR) and electrospray ionization mass spectrometry (ESI-MS) determinations.[25] Confirmation was obtained by X-ray crystallography, where the Cu(I) ion is coordinated by the T-shape N_3 donor set of 2BB at Cu–N distances characteristic of the reduced form of the enzymes. This includes a wide N–Cu–N angle of 165° between the benzimidazole donors, and an average of 97.5° between the benzimidazoles and the central amine (Figure 6). For the previously described cupric analog [(2BB)Cu(H_2O)_2](OTf)_2, speciation studies in aqueous solution revealed a stability constant of log $K = 12.44$, with the dicationic species in the form of $[(2BB)Cu(H_2O)_n]^{2+}$ ($n = 1, 2$) dominating from pH 4 to 8. At higher pH values, a deprotonated species would be present in the form of a cupric hydroxo dimer $[(2BB)Cu(H_2O)_{n-1}]_2(\mu\text{-OH})_2$, see Figure 6.

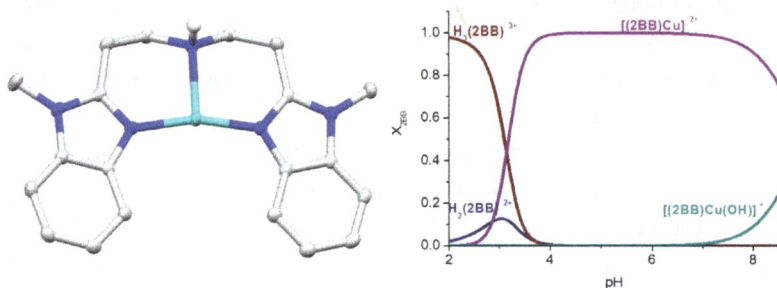

Figure 6. Left: Mercury diagram of [(2BB)Cu]OTf at the 50% probability level; H atoms (except H3 on the central amine) and triflate counterion omitted for clarity. Color code: C, gray; N, blue; O, red; Cu, turquoise. Right: species distribution for the aqueous 2BB/Cu^{II} system.

Deprotonation of $[(\textbf{2BB})Cu(H_2O)_n]^{2+}$ leading to the formation of a Cu^{II}–OH species is relevant to LPMO and model systems: formulations of alternative reactive copper-oxygen species contemplate a deprotonated amine as anionic donor in the active site of LPMOs.[26,27] However, comparison of the pK_a values of Cu^{II}-coordinated water *versus* primary amine favors deprotonation of the former, while access to amido donors under the slightly acidic conditions for optimum LPMO catalytic activity appears unlikely.[28]

Electrochemical determinations by cyclic voltammetry (CV) established that the redox potential $(E_{1/2})$ for the Cu^{II}/Cu^{I} couple of $[(\textbf{2BB})Cu(H_2O)_2](OTf)_2/[(\textbf{2BB})Cu]OTf$ in aqueous solution from pH 4 to 7 has a value of 226 mV *versus* the standard hydrogen electrode (SHE). This value falls within those reported for LPMOs, which range from 150 to 370 mV.[4,10,11,29] Thus, from an electrochemical perspective, the electronic environment provided by **2BB** mimics that of the active sites, and this is further reflected in the reactivity with cellobiose as model substrate. The use of the soluble disaccharide allowed to test the reactivity of the $[(\textbf{2BB})Cu(H_2O)_2](OTf)_2/[(\textbf{2BB})Cu]OTf$ system in the presence of O_2, KO_2, and H_2O_2/triethylamine (Et_3N) as oxidants, instead of the insoluble natural substrate cellulose. Analysis of the reaction mixtures by high-performance liquid chromatography mass spectrometry (HPLC-MS) revealed in all cases the formation of gluconic acid as the product of oxidative cleavage of the disaccharide, together with a product assigned as doubly oxidized glucose, see Figure 7.

Reactions of $[(\textbf{2BB})Cu]OTf$ with dioxygen in different solvents at low temperature hinted at a copper–oxygen species that gives rise to an absorption band at 360 nm and a *d–d* transition at 680 nm. Both $[(\textbf{2BB})Cu(H_2O)_2](OTf)_2$ and $[(\textbf{2BB})Cu]OTf$ afforded inconclusive observations by several spectroscopic techniques, including UV–vis, EPR, and ESI-MS upon

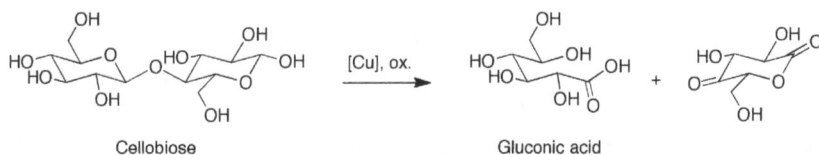

Cellobiose   [Cu], ox.   Gluconic acid

Figure 7. Oxidative cellobiose cleavage with $[(\textbf{2BB})Cu(H_2O)_2](OTf)_2$ or $[(\textbf{2BB})Cu]OTf$ as catalysts and O_2, KO_2, or H_2O_2/Et_3N as oxidants.

addition of KO_2. Finally, evidence for the formation of a peroxodicopper(II) species was obtained from the reaction of the Cu^{II} complex [(**2BB**) $Cu(H_2O)_2](OTf)_2$ with excess H_2O_2/Et_3N in acetonitrile solution at $-30°C$, formulated as $\{[(\textbf{2BB})Cu^{II}]_2(\mu\text{-}\eta^2\text{:}\eta^2\text{-}O_2)\}^{2+}$ based on the Ligand to Metal Charge Transfer band (LMCT) at 365 nm ($\varepsilon \gtrsim 8000$ $M^{-1}cm^{-1}$).[30,31] Cryospray ionization mass spectrometry (CSI-MS) confirmed the presence of a dimeric species assigned as $\{[(\textbf{2BB})Cu^{II}]_2(O_2)\}^{2+}$, although resonance Raman (rR) spectroscopy afforded no isotope-sensitive features under the conditions described earlier for the generation of $\{[(\textbf{2BB})Cu^{II}]_2(\mu\text{-}\eta^2\text{:}\eta^2\text{-}O_2)\}^{2+}$ with either $H_2{}^{16}O_2$ or $H_2{}^{18}O_2/Et_3N$.

Recently, closely related bis(benzimidazolyl)amine-derived copper complexes carried out the oxidative degradation of cellulose.[32] The cupric complexes feature axial EPR spectra, as expected based on their SP geometries in the solid state,[32,33] with very similar parameters among them at $g_\perp = 2.053$ and $g_\parallel = 2.380$ ($A_\parallel = 511$ MHz). Reactions with the H_2O_2/Et_3N oxidant mixture give rise to monocopper species, based on their EPR signals. The most informative UV–vis spectroscopic study with one of the complexes afforded bands at 347 ($\varepsilon = 1280$ $M^{-1}cm^{-1}$) and 610 nm ($\varepsilon = 570$ $M^{-1}cm^{-1}$) in acetonitrile at $-30°C$. These optical transitions are consistent with a hydroperoxo to Cu^{II} LMCT and d-d/LMCT combination bands, based on their intensities and DFT calculations. Preparative scale reactions with cellulose as substrate resulted in up to 67% conversion to the water-soluble products cellobiose, levoglucosan, and aldonic acid, based on the mass recovered of insoluble substrate and HPLC-MS data.

B. Copper(III)/Bis(Carboxamido)Pyridine System

Tolman and coworkers have exploited the dianionic N_3 scaffold provided by the ligand in Figure 8, featuring a central pyridine flanked by two carboxamido donors and sterically-encumbering N-aryl substituents, which enforces meridional coordination.[34] Its potential to mimic the active site of LPMO was identified in 2015 in the form of a Cu^{III}–OH complex (**1**) as oxidant,[35] which is capable of oxidizing strong C–H bonds by hydrogen atom abstraction (HAT). Although the dianionic ligand differs significantly from the neutral donor set provided by the enzymes, its relevance arises from the potential involvement of the invoked deprotonated primary

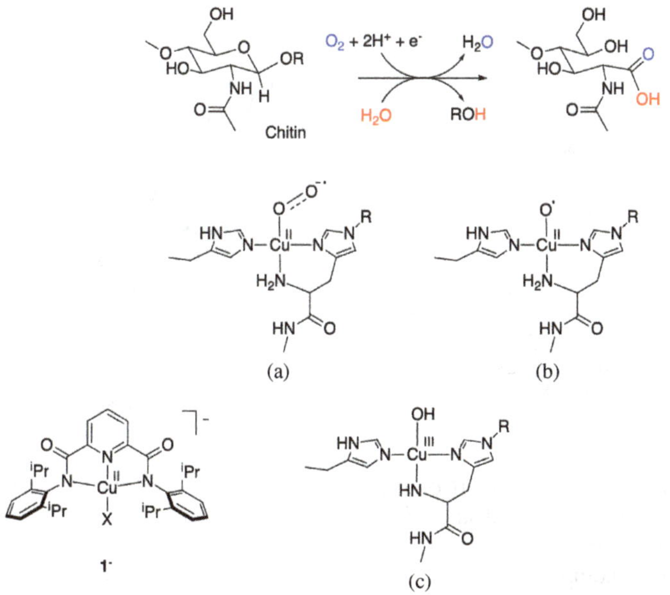

Figure 8. Left: complex **1$^-$**, X = OH. Right: oxidative glycosidic cleavage catalyzed by LPMO and proposed structures for active site oxidants (a–c), including a Cu^{III}–OH complex (c), featuring a deprotonated amine.

amine that is part of the histidine brace in LPMOs.[26,27] Complex **1** was obtained by chemical oxidation of anionic **1$^-$**; the latter features an axial EPR spectrum consistent with tetragonal symmetry at the [Cu^{II}–OH]$^+$ core, and a sharp v(OH) at 3628 cm^{-1} in the IR spectrum.

Solid-state characterization of **1$^-$** was hampered by compositional disorder due to a mixture of chloride and hydroxide ligands at the X position in Figure 8. Nonetheless, the Cu–N1 distance of 1.920(2) Å to the central pyridine and the average of 2.003 Å to the carboxamido N-donors were reliably established.[34] This is the opposite trend to what is observed in LPMO active sites, where the central amine has the longest Cu–N bond distance. Recently, a closely related ligand scaffold that stabilizes the formally Cu^{III} species [Cu^{III}–OR]$^{2+}$ (R = H, CH$_2$CF$_3$) with an electron-donating methoxy group in *para* position relative to the pyridine N-atom allowed its structural authentication, as shown in Figure 9.[36] As expected for the Cu^{III}–OH complex, all Cu–N/O distances shorten

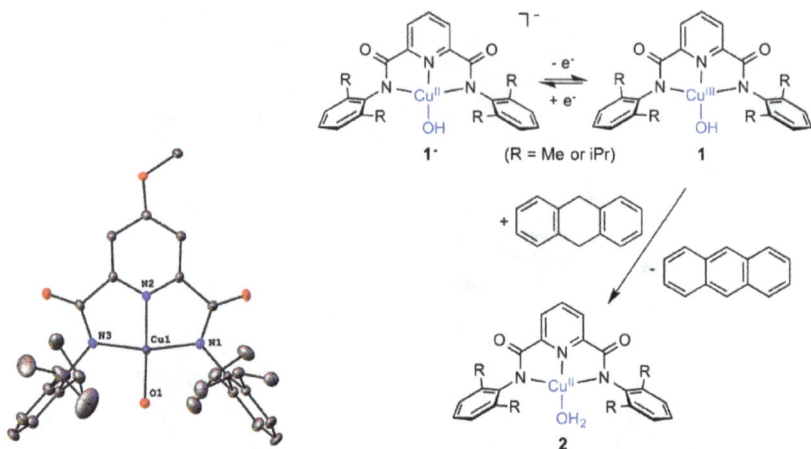

Figure 9. (a) Representation of the solid-state structure at the 50% probability level of a *p*-methoxy substituted ligand supporting a [Cu^{III}–OH]$^{2+}$ complex.[36] (b) H-abstraction by **1** from dihydroanthracene to generate anthracene and [Cu^{II}–OH$_2$]$^{2+}$ complex **2**.[35]

relative to its Cu^{II}–OH counterpart by an average of 0.102 Å. Assignment of **1** as a Cu^{III} species was confirmed by Cu K-edge X-ray absorption spectroscopy (XAS), with a pre-edge feature at an energy that is ~1.7 eV higher relative to its Cu^{II} counterpart **1$^-$**. Support for a diamagnetic species was obtained by ^{1}H NMR spectroscopy and a silent EPR spectrum at X-band frequency.

Conversion of **1$^-$** to **1** was probed by CV, exhibiting a pseudo-reversible oxidation at $E_{1/2}$ = −74 mV for the Cu^{III}/Cu^{II} couple relative to the ferrocenium/ferrocene (Fc$^+$/Fc) couple in THF solution.[35] Its optical spectrum is characterized by a strong absorption band at λ = 540 nm (ε ~12,400 M^{-1} cm^{-1}). Regarding its reactivity, the authors highlight the importance of **1** in the activation of the strong C–H bond of THF that is in the α position relative to an ether functionality, which is related to the glycoside cleavage catalyzed by LPMOs. H-abstraction was tested with substrates that span a range of ca. 20 kcal mol^{-1}, from dihydroanthracene to cyclohexane with the strongest C–H bond at around 100 kcal mol^{-1};[37] in all cases, the reaction is driven by the formation of the O–H bond, as exemplified by the [Cu^{II}–OH$_2$]$^{2+}$ complex **2** in Figure 9. Reactions were monitored by UV–vis spectroscopy, following the decay of the absorption band of **1** at 563 nm in difluorobenzene at −25°C. Intermediate (c), in Figure 8, has some resemblance to **1** due to the presence of an anionic donor, and it has been proposed that it is stabilized by deprotonation of the central amine, although as mentioned before the pK_a value of a primary amine would make this unlikely in the enzymatic system in aqueous solution.[28]

C. Copper(II)/Copper(I)/(Pyridyl,Imidazolyl)Amine System

The complexes reported by Concia and coworkers feature closely related ligands, where the central nitrogen donor takes the shape of an amine (L^{AM}) or an imine (L^{IM}).[38] In both cases, a 2-pyridyl fragment is connected to the central N-atom by an ethylene linker that results in six-membered chelates upon binding of copper. On the other hand, the 2-imidazolyl moiety is connected by a single carbon atom bridge to the central nitrogen, resulting in five-membered chelates in both L^{AM} and L^{IM} (Figure 10).

Structural characterization of the corresponding copper complexes resulted in *pseudo*-octahedral geometries, with both chelating ligands acting as equatorial N_3-donors complemented by a molecule of water, and the perchlorate anions as weak axial O-donors. The Cu–N distances range from 1.97 to 2.03 Å, also within the range of reported LPMO active sites. Aqueous EPR spectra are consistent with axial geometry in solution and a dx^2-y^2 ground state, $g_\perp = 2.059$, $g_\parallel = 2.260$ ($A_\parallel = 530$ MHz) for [L^{AM}Cu(H$_2$O)$_2$](ClO$_4$)$_2$ (**1**), and $g_\perp = 2.060$, $g_\parallel = 2.265$ ($A_\parallel = 530$ MHz) for [L^{IM}Cu(H$_2$O)$_2$](ClO$_4$)$_2$ (**2**). All parameters are comparable to those reported for LPMOs.[4,9–11,29] An important physicochemical property that appears to be directly related to the reactivity of LPMOs for these types

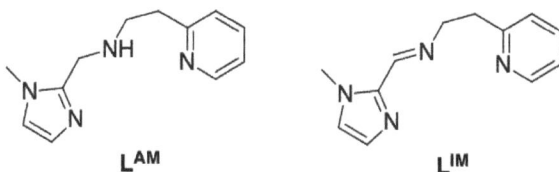

L^{AM} L^{IM}

Figure 10.   Ligands L^{AM} and L^{IM}

of systems is the redox potential, which was determined for both complexes **1** and **2** by CV and redox titrations. The values determined range from 5 to 50 mV *versus* SHE, and contrast with those reported for the metalloenzymes.[4,10,11,29]

Complexes **1** and **2** represent functional models of LPMO on the model substrate *p*-nitrophenyl-β-D-glucopyranoside that features the *p*-nitrophenolate anion as leaving group. Thus, incubation of the complexes with hydrogen peroxide as oxidant at pH 10.5 in carbonate buffer resulted in *p*-nitrophenolate cleavage from the model substrate. The oxidative nature of the reaction was confirmed by the low conversion in the absence of H_2O_2, and by identification of gluconic acid as the product of the initially formed gluconolactone by ESI-MS, see Figure 11. For both **1** and **2**, analysis of the reaction with H_2O_2 in the presence of Et_3N as base in aqueous solution by UV–vis spectroscopy resulted in the detection of an intense band at 305 nm, with a shoulder around 375 nm (Figure 12). The intermediates were postulated as Cu^{II}–OOH species based on previous reports,[39] the *d–d* band at 650 nm, and X-band EPR spectra that resemble those of the parent monometallic complexes.

D. Copper(II)/(Pyridyl)Diazepane Complexes

Mayilmurugan and coworkers devised the pyridyl-appended ligands L1 through L3 from the 1,4-diazepane backbone (1-methyl homopiperazine) for mimicking of the N_3 coordination environment in LPMOs.[40] While the diazepane moiety provides a rigid framework for Cu^{II} binding, there are differences among the ligands regarding the nature of the chelates formed with the pyridyl substituents. L1 features a methylene linker between the unsubstituted 2-pyridyl group and the diazepane N-atom, L2 has an

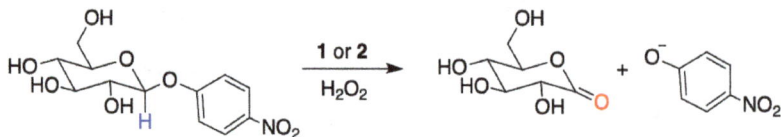

Figure 11. Oxidative cleavage of *p*-nitrophenyl-β-D-glucopyranoside with **1** and **2** as catalysts and H_2O_2 as an oxidant in aqueous buffer solution.

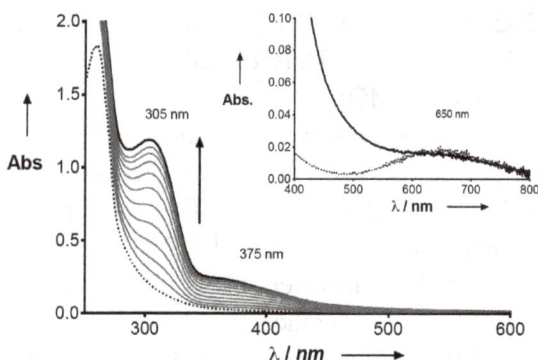

Figure 12. Room temperature absorption spectra of 0.25 mM aqueous buffer solution of [L^{AM}Cu(H$_2$O)$_2$](ClO$_4$)$_2$ (dotted line) and after addition of 25 mM H$_2$O$_2$ (solid lines).

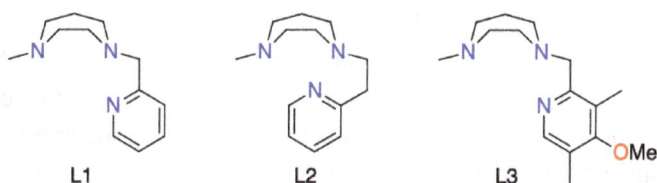

Figure 13. Schematic representation of ligands L1–L3.

ethylene linker, and L3 is characterized by a methylene linker and a 3,5-dimethyl-4-methoxy substitution pattern, see Figure 13. This results in five-, six-, and five-member chelates for L1–L3.

The corresponding cupric complexes [Cu(Ln)(H$_2$O)ClO$_4$](ClO$_4$) (Ln = L1–L3, complexes **1**–**3**) were characterized by standard techniques, and in the case of **1** and **2** by X-ray crystallography. In both cases, distorted SP structures around the Cu^{II} centers are defined at the basal plane by the ligand N$_3$-donors and a molecule of water, and a weak perchlorate O-donor in the axial position. The distortion in **2** is more pronounced, as expected for a more flexible ligand with an ethylene linker defining a six-membered chelate. The Cu–N distances range from 1.99 to 2.02 Å, all similar to those in LPMOs, with the distortion in **2** reflected in the overall longer Cu–N bond lengths as well. EPR spectra acquired in methanol/DMF mixtures are consistent with axial geometry and a dx^2–y^2 ground state, $g_\perp$ = 2.02–2.08, $g_\parallel$ = 2.25–2.37 ($A_\parallel$ = 468–528 MHz); $A_\perp$ were also reported from 423 to 462 MHz. Thus, the EPR parameters for **2** and **3** are comparable to those of LPMOs.[4,9–11,29]

CV studies of **1**–**3** in aqueous solution evidenced quasi-reversible Cu^{II}/Cu^I redox processes, with differences between the anodic and cathodic peaks ΔE from 145 to 163 mV. The redox potentials reported have increasing values at $E_{1/2}$ = 8, 41, and 112 mV relative to the normal hydrogen electrode (NHE). However, this is inconsistent with the values reported *versus* Ag/AgCl as reference, which correspond to the descending values −213, −246, and −317 mV for **1**–**3**, respectively. Thus, further analysis based on the structure of the complexes and ligands is not possible, as is a comparison of the values reported here with those of other model systems and LPMOs.

With these systems, *p*-nitrophenyl-β-D-glucopyranoside was also used as a model substrate, with 30% aqueous hydrogen peroxide and Et_3N as oxidant mixture. Complex **3** featuring electron-donating methyl and methoxy groups on the pyridine donor exhibited the highest rate constant for *p*-nitrophenolate cleavage, as evidenced by UV–vis spectroscopy. The authors attribute this behavior to stabilization of the putative Cu^{II}–OOH intermediate, but this is counterintuitive as it would result in a less electrophilic species. An oxidative process is assumed to be operative, although in preparative scale reactions the co-product was identified as D-allose, see Figure 14. This result contrasts with the observation of gluconic acid as an oxidation product,[25,38] whereas D-allose may be expected in the case of hydrolysis. The reasonable doubt regarding the proposed oxidative nature of the cleavage reaction is further supported by the lack of activity of the most active catalyst **3** in the attempted oxidations of benzyl alcohol and 1-phenylethanol.

Attempts to identify the cupric hydroperoxo intermediate with 10 equivs. of H_2O_2 in the presence of Et_3N in aqueous solution by UV–vis spectroscopy resulted in the detection of LMCT bands around 375 nm, accompanied by *d*–*d* transitions at ca. 630 nm; unfortunately, the extinction coefficient for the LMCT bands was not reported. Addition of *tert*-butyl hydroperoxide to solutions of **3** resulted in a band assigned as LMCT at 380 nm, although the extinction coefficient reported is low for such assignment (ε = 182 $M^{-1}cm^{-1}$). In the latter experiment, the authors

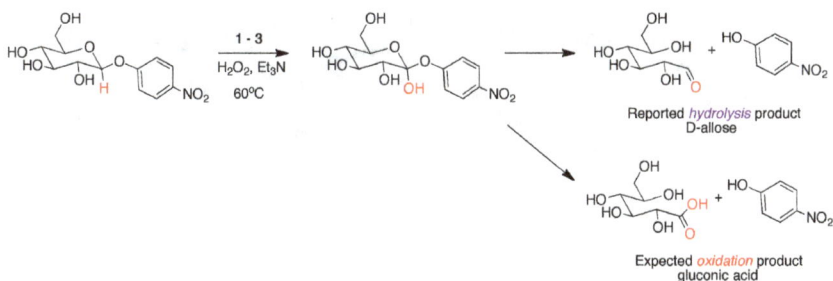

Figure 14. Proposed reaction scheme for the cleavage of *p*-nitrophenol from *p*-nitrophenyl-*β*-D-glucopyranoside with complexes **1–3** as catalysts, with D-allose as co-product.

measured solution FT-IR spectra of the species formed, and determined Cu–O and O–O stretching frequencies for the proposed Cu(II)-OO^{t}Bu species at 750 and 925 cm^{-1}.

E. Copper(II)/Bis(Imidazolyl)Amine System

Modeling of the histidine brace of LPMO by the group of Itoh involved a tridentate N_3 donor that incorporates two imidazole groups and a central amine.[41] In addition to the T-shaped geometry, the authors deemed necessary to have a large "twist" angle between the two imidazole planes, as is the case in the active site of the enzymes, and acidic N–H protons at the same N-heterocycles. These considerations notwithstanding, *τ*-N-methylation of imidazole moieties in N-terminal histidines of fungal LPMOs is a common occurrence.[5,42] The ligand (N-(2-(1H-imidazol-4-yl)-benzyl)histamine) (**LH₃**) and its cupric complex [Cu(**LH₃**)(TFA)₂] (TFA = trifluoroacetate) were thus prepared. The authors mention that under neutral or basic conditions, insoluble blue crystals were obtained when Cu(ClO₄)₂ was added to solutions of the ligand, which they attribute to deprotonation of the N–H group of the imidazolyl moieties and subsequent formation of coordination polymers, but potentiometric measurements were not undertaken to determine their pK_a values.

Structural characterization of [Cu(**LH₃**)(TFA)₂] revealed a SP geometry ($τ_5 = 0.06$),[43] with the basal plane defined by the N_3-donor set and a monodentate TFA counterion; the axial position is occupied by the second

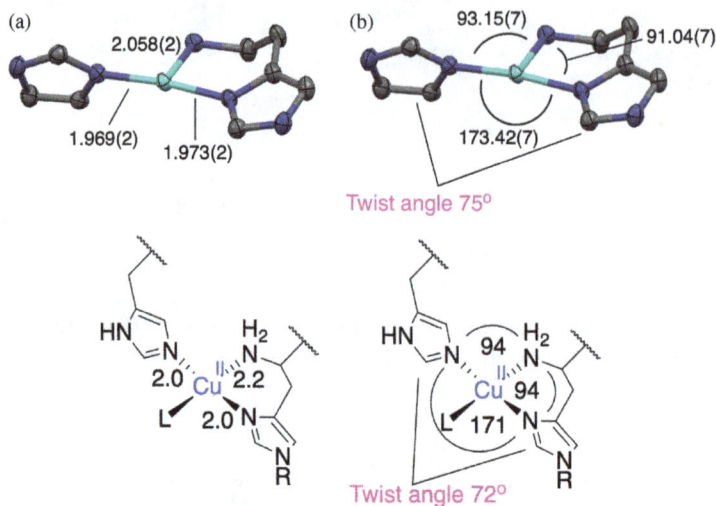

Figure 15. Comparison of (a) bond lengths (Å) and (b) bond angles (°) of [Cu(LH$_3$)(TFA)$_2$] (top) with those of oxidized LPMOs (bottom).

TFA O-atom. The corresponding Cu–N/O distances are similar to those observed in LPMO active sites. As mentioned previously, the authors highlight the angle between the two imidazole planes at 75°, which is close to the average value of 72° determined for the active site of Cu^{II} LPMOs, see Figure 15. Spectral characterization by UV–vis and EPR spectroscopy revealed physicochemical parameters that are comparable to those of the enzymes: a d–d absorption band at λ_{max} = 662 nm (ε = 83 M^{-1}cm^{-1}) typical of SP cupric complexes, and an axial EPR signal characteristic of tetragonal Cu^{II}, $g_\perp$ = 2.062, $g_\parallel$ = 2.266 (A$_\perp$ = 35, A$_\parallel$ = 445 MHz). Electrochemical studies afforded a Cu^{II}/Cu^{I} half wave potential of 323 mV *versus* SHE, with quasi-reversible behavior. The authors stress out that a large "twist" angle between imidazole donors is required to achieve $E_{1/2}$ values that are close to those reported for LPMOs, while model systems with small angles exhibit redox potentials that are lower than 100 mV. However, the Cu^{II}/Cu^{I} couple of the [(**2BB**)Cu(H$_2$O)$_2$](OTf)$_2$/[(**2BB**)Cu]OTf system was incorrectly identified as having a redox potential of 76 mV, while the actual reported value was 226 mV *versus* SHE despite the angle being 14.5° in the cupric complex.[25]

Catalytic tests of $[Cu(LH_3)(TFA)_2]$ with hydrogen peroxide (20 mM) as an oxidant in carbonate buffer at pH 10 were also carried out with *p*-nitrophenyl-β-D-glucopyranoside as substrate. As is the case with $[Cu(Ln)(H_2O)ClO_4](ClO_4)$ (Ln = L1–L3, complexes **1–3**) in Section IV.D, an oxidative cleavage was reported, but the products observed correspond to *p*-nitrophenolate and D-allose. Once again, the identification of D-allose would correspond to hydrolysis rather than oxidation, raising questions about the mechanism of model substrate cleavage.

F. Copper(II)/Bis(Picolyl)Amine System

The most recent example of a simple model system that mimics some aspects of LPMO activity is the one reported by Cowan and coworkers.[44] Thus, bis(picolyl)amine copper(II) chloride (**1a**) was employed along with di- and tetranuclear analogs that will not be discussed, because the enzymatic active sites are mononuclear. The preparation of **1a** was previously reported,[45] and its solid-state structure established that the Cu(II) center has a distorted SP coordination environment, with all N-donors in basal positions.[46] Its redox potential was reported relative to the reversible hydrogen electrode (RHE) at a value of $E_{1/2} = 230$ mV,[47] within the values reported for LPMOs.

The report focuses on glycosidase activity by oxidative cleavage of *p*-nitrophenol conjugated glucose, galactose, and related disaccharides as substrates in the presence of ascorbate and H_2O_2 under physiological conditions.[44] Similar to the earlier examples, the cleavage of *p*-nitrophenolate was monitored by UV–vis spectroscopy, starting with ascorbate and dioxygen, which resulted in ~10% activity when compared to hydrogen peroxide as oxidant. The authors indicate that Fenton-like chemistry might be responsible for the observed reactivity, with hydroxy radicals promoting H-abstraction from the saccharides in the initial step of *p*-nitrophenolate cleavage. Unfortunately, the identity of the saccharide fragments after degradation was not established, leaving the open question of whether the observed glycoside cleavage is indeed oxidative in nature.

V. Concluding Remarks

The meridional donor set provided by the N_3 scaffold in the form of the histidine brace and an additional histidine imidazole in LPMOs is a requisite that needs to be fulfilled to access functional model systems. The effect of the angle between the imidazole rings remains to be understood although it does not appear to be crucial to access the range of redox potentials of the Cu^{II}/Cu^{I} couple in the enzymes. Moreover, the half wave potentials may be a first approximation to the electronic environment around the copper centers in the active sites of LPMOs, but this physicochemical parameter does not appear to be a strict requisite to develop mimics, at least when model substrates are considered. Another observation that emerges is that hydrogen peroxide affords far superior results in the degradation of model substrates and cellulose, which is in line with recent studies that point to H_2O_2 as the oxidant in the enzymatic systems, rather than O_2. It may be time to consider renaming the enzymes as Lytic Polysaccharide Peroxygenases.

While the minimum requirements appear to have been laid out for the development of oxidative glycosidase model systems, key points remain to be addressed: can model systems actually degrade the natural substrates cellulose and/or chitin? This is as challenging to explore with model systems due to the insoluble and recalcitrant nature of such polysaccharides, as it is in nature. Thus far, only the bis(benzimidazole)amine-based copper complexes have demonstrated oxidative cleavage activity on cellulose.[32] Another question that remains to be answered is whether the true oxidant in the enzymatic systems is dioxygen or hydrogen peroxide, and simple model complexes may provide relevant information. In this regard, LPMO mimics may be active at oxidatively degrading model and natural substrates, but they may operate through different reaction mechanisms. In some cases, even the oxidative nature of the model substrate degradation needs to be firmly established. Finally, perhaps the most fundamental question that remains open to speculation is the nature of the copper–oxygen intermediate responsible for the activation of the strong C–H bond of insoluble substrates in LPMOs. Initial proposals considered cupric-superoxide $[Cu^{II}-O_2]^+$ species, but recent experimental and theoretical studies tend to favor a copper-oxyl $[Cu^{II}-O]^+$ intermediate for H-abstraction from strong C–H bonds. Model systems have added to the discussion the potential involvement of a high-valent copper(III)-hydroxide $[Cu^{III}-OH]^{2+}$. Fenton-type reactivity may also be invoked, whether it occurs in the proximity of the copper complexes that may generate hydroxy radicals, or in bulk solution after their release. The combined data obtained from LPMOs themselves, and model systems inspired by the enzymes, should provide answers to these fundamental questions in due course.

VI. References

1. Chu, S.; Majumdar, A. *Nature.* **2012**, *488*, 294–303.
2. Sun, Z.; Fridrich, B.; de Santi, A.; Elangovan, S.; Barta, K. *Chem. Rev.* **2018**, *118*, 614–678.
3. Gómez Millán, G.; Hellsten, S.; Llorca, J.; Luque, R.; Sixta, H.; Balu, A. M. *ChemCatChem.* **2019**, *11*, 2022–2042.
4. Vaaje-Kolstad, G.; Westereng, B.; Horn, S. J.; Liu, Z.; Zhai, H.; Sorlie, M.; Eijsink, V. G. H. *Science.* **2010**, *330*, 219–222.
5. Quinlan, R. J.; Sweeney, M. D.; Lo Leggio, L.; Otten, H.; Poulsen, J.-C. N.; Johansen, K. S.; Krogh, K. B. R. M.; Jorgensen, C. I.; Tovborg, M.; Anthonsen, A.; Tryfona, T.; Walter, C. P.; Dupree, P.; Xu, F.; Davies, G. J.; Walton, P. H. *Proc. Natl. Acad. Sci. USA.* **2011**, *108*, 15079–15084.
6. Phillips, C. M.; Beeson IV, W. T.; Cate, J. H.; Marletta, M. A. *ACS Chem. Biol.* **2011**, *6*, 1399–1406.
7. Sheldon, R. A. *ACS Sustainable Chem. Eng.* **2018**, *6*, 4464–4480.
8. Østby, H.; Hansen, L. D.; Horn, S. J.; Eijsink, V. G. H.; Várnai, A. *J. Ind. Microbiol. Biotechnol.* **2020**, *47*, 623–657.
9. Frandsen, K. E. H.; Simmons, T. J.; Dupree, P.; Poulsen, J.-C. N.; Hemsworth, G. R.; Ciano, L.; Johnston, E. M.; Tovborg, M.; Johansen, K. S.; von Freiesleben, P.; Marmuse, L.; Fort, S. E. B.; Cottaz, S.; Driguez, H.; Henrissat, B.; Lenfant, N.; Tuna, F.; Baldansuren, A.; Davies, G. J.; Lo Leggio, L.; Walton, P. H. *Nat. Chem. Biol.* **2016**, *12*, 298–303.
10. Beeson, W. T.; Vu, V. V.; Span, E. A.; Phillips, C. M.; Marletta, M. A. *Annu. Rev. Biochem.* **2015**, *84*, 923–946.
11. Hemsworth, G. R.; Johnston, E. M.; Davies, G. J.; Walton, P. H. *Trends Biotechnol.* **2015**, *33*, 747–761.
12. Lee, J. Y.; Karlin, K. D. *Curr. Opin. Chem. Biol.* **2015**, *25*, 184–193.
13. Kjaergaard, C. H.; Qayyum, M. F.; Wong, S. D.; Xu, F.; Hemsworth, G. R.; Walton, D. J.; Young, N. A.; Davies, G. J.; Walton, P. H.; Johansen, K. S.; Hodgson, K. O.; Hedman, B.; Solomon, E. I. *Proc. Natl. Acad. Sci. USA* **2014**, *111*, 8797–8802.
14. Bissaro, B.; Rohr, A. K.; Muller, G.; Chylenski, P.; Skaugen, M.; Forsberg, Z.; Horn, S. J., Vaaje-Kolstad, G.; Eijsink, V. G. H. *Nat. Chem. Biol.* **2017**, *13*, 1123–1128.

15. Elwell, C. E.; Gagnon, N. L.; Neisen, B. D.; Dhar, D.; Spaeth, A. D.; Yee, G. M.; Tolman, W. B. *Chem. Rev.* **2017**, *117*, 2059–2107.

16. Striegler, S.; Dunaway, N. A.; Gichinga, M. G.; Barnett, J. D.; Nelson, A.-G. D. *Inorg. Chem.* **2010**, *49*, 2639–2648.

17. Striegler, S.; Barnett, J. D.; Dunaway, N. A. *ACS Catal.* **2012**, *2*, 50–55.

18. Haldar, S.; Patra, A.; Bera, M. *RSC Adv.* **2014**, *4*, 62851–62861.

19. Striegler, S.; Fan, Q.-H.; Rath, N. P. *J. Catal.* **2016**, *338*, 349–364.

20. Comba, P.; Eisenschmidt, A.; Kipper, N.; Schiebl, J. *J. Inorg. Boiochem.* **2016**, *159*, 70–75.

21. Li, Z.; Chen, C. H.; Liu, T.; Mathrubootham, V.; Hegg, E. L., Hodge, D. B. *Biotechnol. Bioeng.* **2012**, *110*, 1078–1086.

22. Castillo, I.; Neira, A. C.; Nordlander, E.; Zeglio, E. *Inorg. Chim. Acta* **2014**, *422*, 152–157.

23. Ambundo, E. A.; Deydier, M.-V.; Grall, A. J.; Aguera-Vega, N.; Dressel, L. T.; Cooper, T. H.; Heeg, M. J.; Ochrymowycz, L. A.; Rorabacher, D. B. *Inorg. Chem.* **1999**, *38*, 4233–4242.

24. Hemsworth, G. R.; Henrissat, B.; Davies, G. J.; Walton, P. H. *Nat. Chem. Biol.* **2014**, *10*, 122–126.

25. Neira, A. C.; Martínez-Alanis, P. R.; Aullón, G.; Flores-Alamo, M.; Zerón, P.; Company, A.; Chen, J.; Kasper, J. B.; Browne, W. R.; Nordlander, E.; Castillo, I. *ACS Omega.* **2019**, *4*, 10729–10740.

26. Bacik, J.-P.; Mekasha, S.; Forsberg, Z.; Kovalevsky, A. Y.; Vaaje-Kolstad, G.; Eijsink, V. G. H.; Nix, J. C.; Coates, L.; Cuneo, M. J.; Unkefer, C. J.; Chen, J. C. *Biochemistry.* **2017**, *56*, 2529–2532.

27. Caldararu, O.; Oksanen, E.; Ryde, U.; Hedegård, E. D. *Chem. Sci.* **2019**, *10*, 576–586.

28. Kandemir, B.; Kubie, L.; Guo, Y.; Sheldon, B.; Bren, K. L. *Inorg. Chem.* **2016**, *55*, 1355–1357.

29. Garajova, S.; Mathieu, Y.; Beccia, M. R.; Bennati-Granier, C.; Biaso, F.; Fanuel, M.; Ropartz, D.; Guigliarelli, B.; Record, E.; Rogniaux, H.; Henrissat, B.; Berrin, J.-G. *Sci. Rep.* **2016**, *6*, 28276.

30. Casella, L.; Gullotti, M.; Radaelli, R.; Di Gennaro, P. *J. Chem. Soc., Chem. Commun.* **1991**, 1161–1612.

31. Baldwin, M. J.; Root, D. E.; Pate, J. E.; Fujisawa, K.; Kitajima, N.; Solomon, E. I. *J. Am. Chem. Soc.* **1992**, *114*, 10421–10431.

32. Castillo, I.; Torres-Flores, A. P.; Abad-Aguilar, D. F.; Berlanga-Vázquez, A.; Orio, M.; Martínez-Otero, D. *ChemCatChem.* **2021**, *13*, 4700–4704.

33. Rodríguez Solano, L. A.; Aguiñiga, I.; López Ortiz, M.; Tiburcio, R.; Luviano, A.; Regla, I.; Santiago-Osorio, E.; Ugalde-Saldívar, V. M.; Toscano, R. A.; Castillo, I. *Eur. J. Inorg. Chem.* **2011**, 3454–3460.

34. Donoghue, P. J.; Tehranchi, J.; Cramer, C. J.; Sarangi, R.; Solomon, E. I.; Tolman, W. B. *J. Am. Chem. Soc.* **2011**, *133*, 17602–17605.

35. Dhar, D.; Tolman, W. B. *J. Am. Chem. Soc.* **2015**, *137*, 1322–1329.
36. Krishnan, V. M.; Shopov, D. Y.; Bouchey, C. J.; Bailey, W. D.; Parveen, R.; Vlaisavljevich, B.; Tolman, W. B. *J. Am. Chem. Soc.* **2021**, *143*, 3295–3299.
37. Luo, Y. *Handbook of bond dissociation energies*, 1st Ed. CRC Press, USA, 2002.
38. Concia, A. L.; Beccia, M. R.; Orio, M.; Ferre, F. T.; Scarpellini, M.; Biaso, F.; Guigliarelli, B.; Réglier, M.; Simaan, A. J. *Inorg. Chem.* **2017**, *56*, 1023–1026.
39. Wada, A.; Harata, M.; Hasegawa, K.; Jitsukawa, K.; Masuda, H.; Mukai, M.; Kitagawa, T.; Einaga, H. *Angew. Chem. Int. Ed.* **1998**, *37*, 798–799.
40. Muthuramalingam, S.; Maheshwaran, D.; Velusamy, M.; Mayilmurugan, R. *J. Catal.* **2019**, *372*, 352–361.
41. Fukatsu, A.; Morimoto, Y.; Sugimoto, H.; Itoh, S. *Chem. Commun.* **2020**, *56*, 5123–5126.
42. Hemsworth, G. R.; Taylor, E. J.; Kim, R. Q.; Gregory, R. C.; Lewis, S. J.; Turkenburg, J. P., Parkin, A.; Davies, G. J., Walton, P. H. *J. Am. Chem. Soc.* **2013**, *135*, 6069–6077.
43. Addison, A. W.; Rao, T. N.; Reedijk, J.; van Rijn, J.; Verschoor, G. C. *J. Chem. Soc. Dalton Trans.* **1984**, 1349–1356.
44. Yu, Z.; Thompson, Z.; Behnke, S. L.; Fenk, K. D.; Huang, D.; Shafaat, H. S.; Cowan, J. A. *Inorg. Chem.* **2020**, *59*, 11218–11222.
45. Humphryes, K. J.; Karlin, K. D.; Rokita, S. E. *J. Am. Chem. Soc.* **2002**, *124*, 8055–8066.
46. Choi, K.-Y.; Ryu, H.; Sung, N.-D.; Suh, M. *J. Chem. Crystallogr.* **2003**, *33*, 947–950.
47. Tse, E. C. M.; Schilter, D.; Gray, D. L.; Rauchfuss, T. B.; Gewirth, A. A. *Inorg. Chem.* **2014**, *53*, 8505–8516.

7 Structure and Function of Cu–Peptoid Complexes

Anastasia E. Behar,* Pritam Ghosh,* and Galia Maayan*,†

*Schulich Faculty of Chemistry, Technion — Israel Institute of Technology, Technion City, 3200008 Haifa, Israel
†gm92@technion.ac.il

The remarkable efficiency and selectivity of natural biopolymers in carrying out valuable functions such as molecular recognition and catalysis have been a source of inspiration for the design and synthesis of functional chemical systems. Mimicking the capabilities of biopolymers requires the ability to imitate their unique properties, such as their well-defined three-dimensional structures and their high selectivity toward specific metal ions that are required for their utility, as well as to imitate conceptual features of their activity, such as intramolecular cooperativity. A promising approach to achieve biomimetic properties involves (i) facile synthesis of a versatile oligomeric backbone that enables the incorporation of several functional groups with sequence specificity and (ii) control over structure formation including the ability to enforce folding and self-assembly. These aptitudes should allow generating functional peptidomimetic systems in which various components act cooperatively toward a specific biomimetic function including selective recognition as well as efficient and selective catalysis.

Copper ions are key elements in both the structure and function of natural biopolymers, being employed as cofactors in many essential proteins,[1] facilitating folding or stabilizing specific conformations, and regulating vital functions in living organisms,[2] including oxygen transport, cellular stress response, antioxidant defense, and more. In addition, Cu ions are vital components of many enzymes, for example, the zinc–copper superoxide dismutase.[2] Despite the great importance of Cu within biological systems, its overloading can be potentially toxic to living cells and may cause oxidative stress, neurodegenerative diseases including Alzheimer's disease,[3] and other disorders such as Wilson's disease,[4] and cancer.[4,5] A possible treatment of Cu intoxication is *via* the administration of chelating molecules (chelation therapy),[6,7] designed for the specific binding of Cu ions.

In this chapter, we will describe peptide mimics called peptoids — N-substituted glycine oligomers — that can either bind Cu ions with high specificity or fold, change their conformation, and/or self-assemble *via* Cu coordination. We will also present the applications of Cu–peptoid complexes in intramolecular cooperative catalysis, a concept inspired by the activity of enzymes.

I. Introduction

Biomimetic oligomers are synthetic molecules akin to natural biopolymers, namely peptides, proteins, and oligonucleotides.[8–10] Both natural biopolymers and their artificial mimics are sequence-specific oligomers capable of folding into well-defined three-dimensional structures in solution. One difference between them, however, is the identity of their backbone: while peptides and proteins are composed of α-amino acids, biomimetic oligomers are assembled from non-natural monomers.[11,12] Among the peptide mimics, peptoids are unique, as they combine properties of both biological materials and synthetic polymers; they are inert toward many catalytic transformation[13] and oxidative conditions[14–16]; and possess better physiochemical and pharmacokinetic properties relative to α-peptides, including stability in oxidative conditions[13] and high temperatures,[17] tolerance toward high salts concentration and various pH conditions,[13,18] enhanced proteolytic stabilities,[19] and increased cellular permeability.[20] These advantages, taken together with their ease of synthesis on solid support (see the following section), their biomimetic structure, and ability to control their sequence and structure, hold great potential in developing peptoids as biomimetic functional materials including selective chelators for metal ions and catalysts.

A. Peptoid Synthesis

Peptoid oligomers have identical backbone to that of polypeptides, but the side chains are appended to the nitrogen rather than to the α-carbon (Figure 1a). Consequently, peptoids lack backbone chirality and cannot form hydrogen bonds. In addition, the amide bonds in peptoids are tertiary and this allows isomerization between the *cis* and *trans* conformations (Figure 2a),[21] in contrast to the α-peptides, where the *trans* conformation is preferable. Furthermore, the lack of amide protons restricts the stabilization of secondary structures by backbone hydrogen bonding, as observed in α-peptides. Nevertheless, peptoids have several properties that make them superior over peptides, and one of the main advantages is the ease of the peptoids synthesis. Peptoids can be synthesized efficiently on solid support, using the submonomer approach[22] (Figure 1b), which involves two main steps: N, N-diisopropylcarbodiimide (DIC) mediated

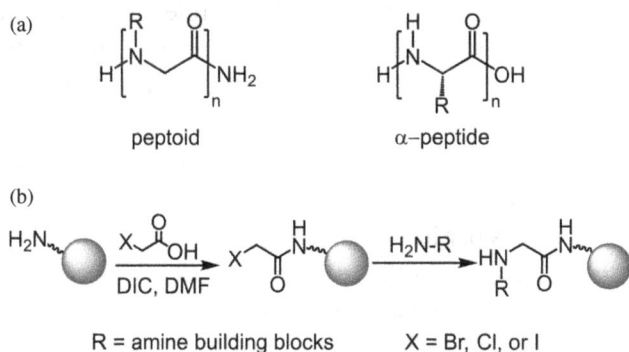

Figure 1. (a) Generic structure of a peptoid compared to α-peptide. (b) Schematic representation of peptoid submonomer synthesis.

acylation with bromo- or chloroacetic acid, followed by amine displacement with a primary amine to form N-substituted glycine. This two-step sequence is repeated iteratively to obtain the desired oligomer, followed by the cleavage of the oligomers from the resin. This synthesis can be performed using any primary amine, which is either commercially available or synthesized; the synthesis typically does not include any protection/deprotection steps, and thus the submonomer approach allows to introduce various functional groups within the peptoid sequence.[23] The submonomer approach makes the construction of peptoids relatively inexpensive compared to the synthesis of α-peptides.

B. Secondary Structure of Peptoid Oligomers

In contrast to peptides, where folding arises from backbone hydrogen bonding interactions, the backbone nitrogen atoms in peptoids are substituted, and therefore lack the ability to form hydrogen bonding. This results in highly flexible backbones and complicates the design of well-defined secondary structures in peptoids. However, different methods were developed to allow folding of peptoids into secondary structures by the incorporation of the bulky chiral side chains within the oligomer sequence, thus restricting backbone conformation.[17] More specifically, incorporation of branching bulky substituents induce steric repulsion between side chains, leading to charge-charge repulsion with the backbone carbonyls; hydrophobic interactions or $n \rightarrow \pi^*$ or π-π interactions between aromatic bulky substituents play a role in peptoid folding.[17] Peptoid oligomers have been found to be able to fold into loop,[17] helix,[17,21,24] or turn.[14,25] Among them, one of the most studied secondary structures of peptoids throughout the last decades is the helix, which typically resembles the polyproline type I (PPI) peptide helix, and depending on the type and number of the bulky chiral side chains, have a pitch of roughly three residues per helical turn.[17] In contrast to peptides, in which the amide bond is mostly in the *trans* conformation, the amide bond of peptoids goes through rapid isomerization between the *cis* and *trans* conformations leading to significant conformational heterogeneity in solution. Nevertheless, a peptoid helical structure could be obtained even in short oligomers (as short as pentamers).[17] Interestingly, among various chiral side chains that were shown to induce helicity within peptoids, a few of them have been determined as the side chains that shift the *cis/trans* isomerization toward a majorly of *cis* amide bonds, for example, *N*-(S)-1-phenylethylglycine (Nspe), *N*-(R)-1-phenylethylglycine (Nrpe) (S)-*N*-(1-naphthylethyl)glycine (Ns1npe)[17,24,26] and some were even able to lead to "*all cis*" helices, for example, (R)-(−)-3,3-dimethyl-2-butylamine (Nr1tbe), (S)-(+)-3,3-dimethyl-2-butylamine (Ns1tbe)[21,27] (Figure 2b). Several sequences requirements for obtaining peptoid helices were established. First, the peptoid helix can be stabilized when the majority of the peptoid's side chains (at least two-third) are bulky and chiral.[28] Second, placing a chiral bulky monomer at the C-terminal of the peptoids sequence can induce folding of the peptoid into a helical structure, so-called "sergeant-soldier" effect; in this case, the overall number of chiral subunits within peptoid scaffold could be reduced.[28] Third, the longer the peptoid sequence is, the more stable is the helical structure (peptoid decamers and beyond).[28]

The secondary structure of the peptoids can be initially estimated by circular dichroism (CD) spectroscopy. For example, the typical spectra of Nspe-bearing peptoids show typical double minima indicating a negative

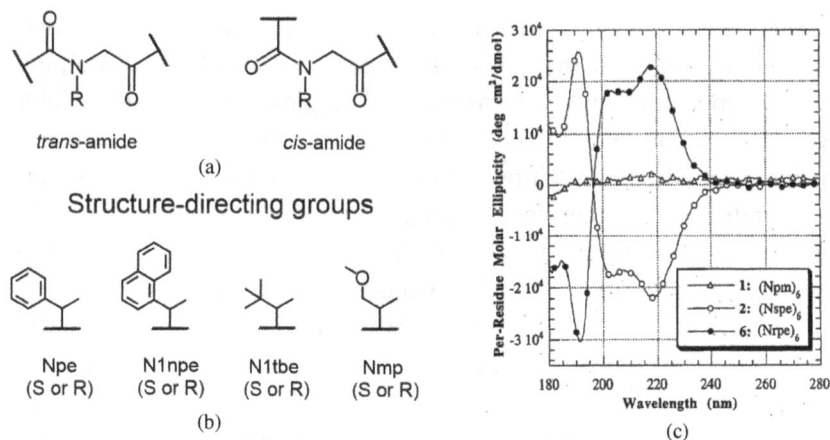

trans-amide   cis-amide
(a)

Structure-directing groups

Npe   N1npe   N1tbe   Nmp
(S or R)   (S or R)   (S or R)   (S or R)
(b)

(c)

Figure 2. (a) Isomerization of amides in the peptoid backbone. (b) Common α-chiral monomers used to induce secondary structure in peptoids. (c) CD spectra of the polypeptoids (*N*spe)₆ and (*N*rpe)₆. Peptoid concentration was 0.2 mg/ml (≈1 mM residue molarity for all species) in 10 mM sodium phosphate (pH 7.0) at 25°C.[28]

ellipticity near 218 nm and 202 nm, and a maximum indicating a positive ellipticity near 190 nm (Figure 2c).[17,28] These bands correspond to $n \rightarrow \pi^*$ transition of the amide chromophore and high- and low-wavelength components of the exciton split $\pi \rightarrow \pi^*$ transition. The similar, mirror-like pattern of opposite chirality (double-maxima and minima, correspondingly) is observed for Nrpe-bearing peptoids (Figure 2c).[17] In accordance with the helical peptoids having a pitch of about three residues per turn,[17] pendant groups located at positions i and $i + 3$ along the peptoid backbone are facing the same side of the helix, and this can serve as a handle for the design of peptoids as selective chelators for metal ions, including Cu.

II. Structural Aspects of Cu(II)–Peptoid Complexes

A. Structural Design of Cu(II)–Peptoid Complexes

Over the last years the main focus of our group has been the synthesis of metal-binding peptoids and their coordination to metal ions, specifically Cu(II). Our main working hypothesis for designing peptoid oligomers with both high affinity and high selectivity to Cu(II), was that the preorganization of specific coordinative ligands at positions i and $i + 3$ of unstructured or structured (helical) oligomers will enhance their affinity to Cu(II) compared with two such ligands tethered together *via* a simple chemical bond or in any other unstructured oligomer, and may also enable secondary structure stabilization. For the generation of helical metal-binding peptoids, bulky chiral side chains (see previous section) were incorporated in the other positions along the sequence, not occupied by the metal-binding ligands, provided that these chiral bulky side chains constitute the majority of the substituents along the peptoid scaffold.

These metal-binding peptoids are expected to form complexes of the type PCu(II) (Figure 3).[29–31] Peptoids bearing two metal-binding sites, either identical (or similar) or different — having a significantly distinct metal affinity that can bind metal ions in different coordination geometries were developed for the selective intermolecular binding of either two Cu(II) ions or one Cu(II) ion and one different metal ion, yielding complexes of the type P_2Cu_2 or P_2CuM, where M = Zn(II), Co(II), or Fe(III),

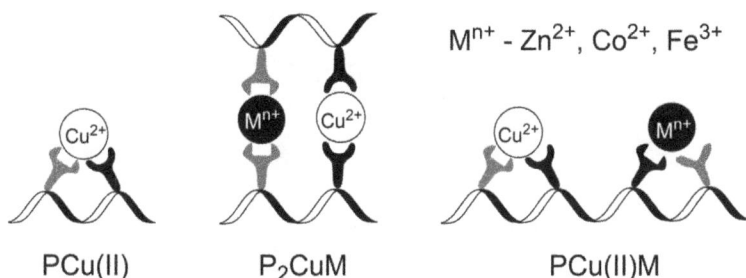

M^{n+} - Zn^{2+}, Co^{2+}, Fe^{3+}

PCu(II) P$_2$CuM PCu(II)M

Figure 3. Generic representation of the different types of the metallopeptoid complexes structures.

respectively[29,30] (Figure 3). Controlling the interactions between these two ligands *via* metal coordination could be further extended by placing them at i and $i + n$ positions where $n \neq 3$. Taking this approach, we have shown that it is possible to separate between two or more different coordination sites and design oligomers capable of binding two ions, being either Cu(II) ions or Cu(II) ion and another metal ion. The latter binding was shown to proceed selectively in distinct sites forming various hetero bimetallic peptoid complexes PCu(II)M (Figure 3).[32] The use of peptoid sequences as scaffolds for coordinative ligands enabled to tune their chemical properties, such as lipophilicity and chirality, simply by modifying the non-coordinative side chains, thus controlling the peptoids' compatibility with various applications.[30,31]

B. Choice of the Metal-Binding Ligands

The coordination properties of the incorporated ligands were also considered in the design of the new selective metal-binding peptoids; two ligands that are able to provide preferable coordination geometry to Cu(II), for example, penta-coordination, were placed at positions i and $i + 3$ to enable strong binding in an intramolecular fashion. According to Hard and Soft Acids and Bases (HSAB) principle,[33] Cu(II) is considered borderline hard/soft metal ion; thus in nature, protein–copper(II) binding sites are dominated by side chains containing borderline ligands that combine coordination the atoms N and S and O.[34] In addition, based on Ligand Field Theory (LFT) and the ligand field stabilization energy (LFSE),[35]

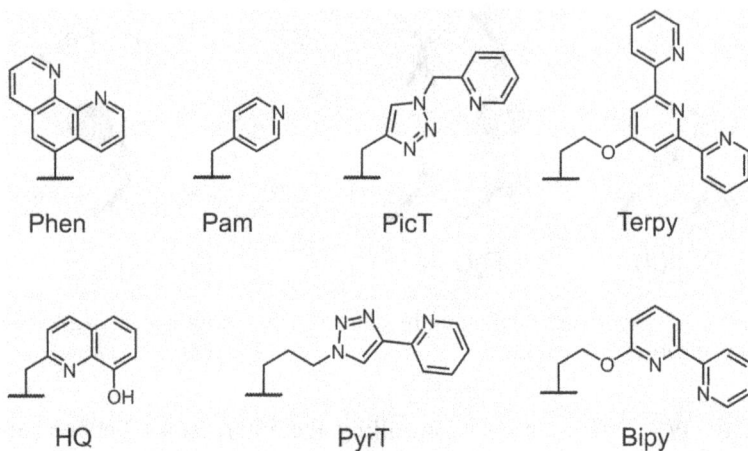

Figure 4. Metal-binding ligands for Cu(II) used in peptoids design.

Cu(II) is a d^9 system, which exhibits geometric preferences based in part on LFSE, and is found to be coordinated by 4, 5, or 6 ligands in square planar, square pyramidal, or axially distorted octahedral geometries. Considering preferences from both HSAB and LFT, versatile examples of metal-binding ligands for Cu(II) ions that were incorporated within peptoid sequences are depicted in Figure 4. These mono-, bi-, and tridentate ligands combine both N and O atoms and therefore are varied in their electronic properties and in their possible geometrical coordination modes thus enable the formation of either tetragonal-, pentagonal-, or hexagonal-coordinated complexes. All these ligands are also known to bind Cu(II) ions with high affinity (stability constants of Cu(II) complexes Phen, Pam, Bipy, Terpy, HQ, and PicT are logK = 17.9, 10.2, 17.85, 6.97, 13.00, and 7.33 respectively, as obtained by potentiometric titrations at 25°C).[36–40]

Metal-binding (MB) ligands are introduced into the peptoid backbone unprotected (thus avoiding additional protection and de-protection synthetic steps), in the amine displacement step as a primary amine, which are readily available — Phen and Pam, or could be synthesized according to procedures developed by our group: HQ ligand has been previously synthesized by a one-step hydrogenation reaction[41] and Terpy and Bipy

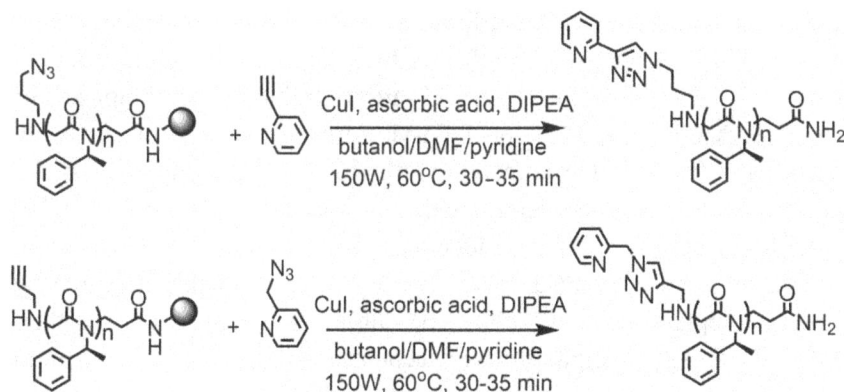

Figure 5. Solid-phase synthesis of the peptoid oligomers by the submonomer method and their copper(I)-catalyzed azide–alkyne [3 + 2] cycloaddition (click) reaction.

by one-step nucleophilic substitution reaction.[27,41] In addition, inclusion of pendant functional groups within peptoid sequences can be also accomplished by the Cu(I)-catalyzed azide-alkyne (3 + 2) cycloaddition reaction ("click") on solid support by microwave irradiation at 60°C. (Figure 5). By using this approach, we have recently exploited this synthetic path for incorporating the pyridine-triazole ligands PyrT and PicT within four helical peptoids.[42]

C. Characterization of Cu(II)–Peptoid Complexes

1. Evaluation of Cu(II) binding and synthesis of Cu(II)–peptoid complexes

The coordination capabilities of metal-binding peptoids to Cu(II) can be evaluated using several analytical techniques. Initial assessment of Cu(II) binding to the peptoid ligand(s) is usually done in solution *via* titrations of Cu(II) solutions in known concentrations (in solvents such as acetonitrile, water, and methanol, where both the Cu(II) salts and the peptoids are soluble) to peptoids solutions in known concentrations, followed by UV–vis spectroscopy. Based on the obtained data, the Cu(II):peptoid stoichiometric ratio is determined from ratio plots and/or Job plots. Further confirmation of the stoichiometric ratio can be obtained from mass spectrometry (MS) techniques, specifically by isotopic analysis that can support a 1:1, 1:2, or 2:1 Cu(II):peptoid ratio (intramolecular complexes of the type PCu/PCuM, PCu$_2$ or P$_2$CuM, respectively; see examples in Figure 6)[29–32] or a 2:2 Cu(II):peptoid ratio (intermolecular duplexes of the type P$_2$Cu$_2$).[29] Using the determined ratio, metal complexes are synthesized in a larger scale and their identity is confirmed by MS. Cu–peptoid complexes of the type PCu, PCu$_2$, and P$_2$Cu$_2$ are typically prepared by treating a peptoid solution with Cu(II) salt under stirring, following their isolation either by precipitation with a counter ion or, in the case of neutral complexes, by solvent evaporation and further purification (*e.g.*, recrystallization).[31] Heteronuclear metallopeptoids of the types P$_2$CuM and PCuM are usually prepared in two methods, under kinetic control and under thermodynamic control. Following the first method, metal ions are added step by step such that one metal is added first, stirred for some time, and its binding to the peptoid solution is verified by UV–vis and MS spectroscopies. Subsequently, the second metal is added, and the binding process and analysis is repeated until the entire oligomer is bound.[30–32] Following the second method, a mixture of metal ions is added to the peptoid solution simultaneously, heated and/or stirred for several hours, isolated, and characterized as described above.[31]

Detecting the binding of two different metal ions, namely one Cu(II) ion and one other metal ion, to one peptoid could be also done by UV–vis spectroscopy using the kinetic control approach. First, each metal ion solution is added separately to the peptoid aiming to coordinate at its designed site, and its UV–vis spectra is compared with the one previously found in order to ensure that the binding indeed occurs in the desired site. In addition, as each coordinative ligand exhibits different absorbance band(s), the changes in each band upon the addition of each metal ion solution is followed separately by UV–vis. This provides information about the number and type(s) of ligands involved in the binding of each metal ion. All the acquired data from the kinetic experiments was compared with the data obtained from the binding of two metal ions from their mixture solution. Overlap between different bands is possible, but in most cases enough data is gained from non-overlapping bands, and the overall information allows to precisely determine the coordination of different metal ions to the same peptoid.

2. Determining the association constants and evaluating the selectivity of the peptoid chelators to Cu(II)

One of the important parameters for evaluating the Cu(II) binding to peptoids is the association/dissociation constant of the Cu(II)–peptoid complex. Association constants of metallopeptides can be determined by one of the following methods depending on the solubility of the peptoid and the MB ligands. For peptoids with low water solubility UV–vis titrations in aqueous solutions containing 10–20% methanol (the peptoid is dissolved in methanol prior to water addition). Low peptoid and metal ion(s) concentrations are used in order to ensure low ionic strength. Association constants are estimated by fitting the obtained data *via* non-linear regression using the appropriate kinetic equations.[15] For water-soluble peptoids, isothermal titration calorimetry (ITC) measurements can be used in buffer solutions.[42,43] Association constants and other physical parameters are extrapolated directly, using the ITC software. In addition, in order to verify the results (obtained in either method), competition experiments can be conducted, either between the metal ions competing over the peptoid ligand, or between the peptoid ligand and a strong Cu(II) chelator with a known association constant. Using the first competition method, two different metal ions are titrated with a peptoid solution, followed by UV–vis spectroscopy and MS. From these experiments, quantitative data about the selectivity of one metal ion over the other is deduced. Using the second competition method, the dissociation constant for Cu(II)–peptoid complex could be determined by metal ions titrations with both chelators followed by UV–vis spectroscopy. From the obtained UV–vis data, using the appropriate equations and pH correction factors, dissociation constant is found from the calculated slope.[44]

To determine the selectivity of the peptoids toward Cu(II), several approaches could be applied. Typically, each peptoid is treated with a mixture of different metal ions in different concentration and the solution mixture is investigated. First, the solution mixture is analyzed by UV–vis and MS spectroscopies in order to determine the identity of the complexes. The obtained spectra are compared to the spectra of each metal–peptoid complex that was generated separately (not from a mixture). Identity between the spectra obtained from the mixture solution and the spectra of the Cu(II)–peptoid complex indicates that the peptoid is selective to Cu(II).[31] Second, the association constants of the peptoids with each metal ion from the tested mixture are compared in order to evaluate which ion has the highest affinity to the peptoid; highest affinity to Cu(II) supports selectivity to this ion. Finally, the precipitate formed from the mixture solution is filtered, the filtrate solution is analyzed by inductively coupled plasma optical emission spectroscopy (ICP-OES) technique and the filtrate is analyzed again by UV–vis and MS. The absence of Cu(II) from the filtrate and its sole presence in the precipitate indicates its selective binding and extraction by the peptoid.[31]

3. Structure and coordination sphere of Cu centers within Cu(II)–peptoid complexes

Electron paramagnetic resonance spectroscopy (EPR) and X-ray analysis are usually used to characterize the structure and coordination geometry of Cu(II)–peptoid complexes in solution and in the solid state. The experimental EPR bands of solid or frozen solution samples are recorded in the presence of an internal standard and the obtained spectra are simulated in order to obtain the Hamiltonian parameters (g and $A_\parallel$ values). According to these parameters, the coordination geometry of the Cu(II)–peptoid complexes can be estimated. Typically, the EPR spectra of Cu(II)–peptoids resemble tetra-[30,42,45] or penta-[31,46–49] coordination geometry of the Cu(II) center. For tetra-coordinated Cu(II) complexes, $g_\parallel > g_\perp$ and the quotient $g_\parallel/A_\parallel$ (cm) is calculated.[50] The value of this quotient indicates whether the coordination geometry of the Cu(II) center is square planar (with quotient $g_\parallel/A_\parallel$ range between 105 and 135[42]) or has a tetrahedral distortion (when the quotient $g_\parallel/A_\parallel$ is higher than 135[30,45]). For penta-coordinated Cu(II), typically two g values are obtained, with $g_\parallel > g_\perp > 2.0023$ and $A_\parallel$ been any measurable value, and then the geometry is determined between square-pyramidal or trigonal-bipyramidal according to exact values of these parameters.[13,31,47] In some cases, the determination between tetra- or penta-coordination only on the basis of the EPR Hamiltonian parameters could be rather challenging, and thus should be supported by other analytical methods including X-ray, FT-IR, CD, and UV–vis.[46] Finally, in some cases the intermediate situations exhibiting three g values could be observed, resulting in the so called "rhombic" spectrum; the coordination of Cu(II) center in these cases is intermediate between square-pyramidal and trigonal-bipyramidal.[51] For complexes of this type, a parameter R (where $R = (g_y-g_z)/(g_x-g_y)$ with $g_x > g_y > g_z$) should be calculated in order to determine the predominance of the ground state among two possible, and as a result, the prevalence of one of them could be established. If $R > 1$, then the greater contribution to the ground state arises from d_z^2 (trigonal-bipyramidal), if $R < 1$, the greater contribution to the ground state arises from $d_{x^2-y^2}$ (square-pyramidal).[47,49]

In cases where crystals suitable for single crystal X-ray analysis are obtained, X-ray analysis of the obtained crystals provides precise information about the bond distances and the exact geometry of the Cu(II) metal center, including second coordination sphere and full structure of metallopeptoid in the solid state.[46]

To evaluate the influence of Cu(II) on the secondary structure of the peptoid, CD spectroscopy measurements of both the peptoid and its Cu(II) complex are typically performed in solution. The comparison between the obtained CD spectrum of the peptoid and this of the Cu(II)–peptoid complex indicate whether the peptoid had some secondary structure before Cu(II) binding and if so, whether this structure is stabilized or destabilized upon Cu(II) binding. In case the peptoid was not structured prior to Cu(II) binding, the spectra provide information to whether Cu(II) binding can enforce folding and eventually lead to some secondary structure of the Cu(II)–peptoid.

III. Effect of Cu(II) Binding on the Structure of Peptoid

A. Cu(II) Binding as a New Approach for Peptoid Folding

Since the first introduction of peptoids as a new class of peptidomimetic oligomers 30 years ago, their folding in solution and the requirements for achieving well-defined secondary structures were established and described in details (see section I.B). Folding of peptoids into a stable helical secondary structure in solution generally requires incorporation of chiral bulky side chains within peptoids sequence, such that they occupy at least two thirds of the peptoids length.[17] This strategy, although been well known and commonly used in the field, restraints potential applicability of the peptoid helices: currently, only several side chains are known to enable folding, out of the numerous that exist, and these do not contribute to peptoid functionality as they only serve structural purposes. Furthermore, the helix-inducing side chains should occupy the majority of the peptoid structure, and therefore the introduction of functional groups is limited. Thus, finding ways to control both the structure and function of peptoids is required for the development of unique functional biomimetic materials. Among a variety of interactions known to enforce folding within biopolymers, including hydrogen bonding, solvophobic effects, and metal–ligand interactions, the latter, and particularly in the case of copper, are found to enable folding and function. As peptoid helices resemble polyproline-type helices (PP-I and PP-II),[17] with a helical pitch of three residues per turn, we have suggested that the incorporation of MB ligands in i and $i + 3$ positions of peptoid sequences might enforce peptoid folding and eventually will result in peptoid helices upon metal binding.

The first Cu(II) peptoids were generated from helical peptoids having MB ligands in i and $i + 3$ positions and Nspe side chains (that ensure peptoids helicity) in the other positions along the peptoid sequence.[15,42] These examples showed that Cu(II) binding has a minor effect on peptoids initial secondary structure. On the other hand, Cu(II) binding to HQ-containing peptoids resulted in new CD peaks between 240 nm and 280 nm, the region corresponding to the HQ π-π* transition, reflecting the transmission of stereogenic character of the helical peptoids scaffold to the metal center. Although this was a unique illustration of chiral induction to Cu(II) center in peptoids, it was not determined whether the chirality of the metal centers arises from the peptoid helicity or simply from the fact that these oligomers are chiral.

In order to probe this point and also explore if it is possible to induce folding within unfolded peptoid sequences by metal coordination only, a peptoid heptamer having two HQ groups at positions 3 and 6 (from the

N-terminal), and chiral, non-bulky (S)-1-methoxy-2-propyl (Nsmp) groups in the other five positions was synthesized (**7mer-HQ2**, Figure 6a).[30] The Nsmp groups are not secondary structure inducers, but are hydrophilic, thus **7mer-HQ2** was not expected to have a helical or any other secondary structure but was expected to be water soluble, and it was suggested that folding might be enforced by Cu(II) binding to form water-soluble helical Cu–peptoid.[30] Although both UV–vis and ESI-MS of the peptoid after Cu(II) binding were consistent with the formation of 1:1 intramolecular (**7mer-HQ2**)Cu complex, no helix formation was observed upon binding of Cu(II) to **7mer-HQ2**, as revealed from the CD; the spectra shape in the range of 190–230 nm for both, metal-free peptoid **7mer-HQ2** and its Cu(II) complexes, suggested completely disordered structures. Consequently, the peptoid dodecamer **12mer-HQ4**, bearing four HQ ligands that form two MB sites at positions i and $i + 3$ (Figure 6b) for the formation of an intramolecular complex, of the type PCu$_2$, was synthesized and treated with Cu(II) ions. The formation of the complex (**12mer-HQ4**)Cu$_2$ was verified by UVvis and ESI-MS (Figure 6b). This time, the CD spectra of **12mer-HQ4** after addition of Cu(II) ions revealed a simultaneous, although minor, decrease in the CD magnitude at 191 nm and increase near 210 nm, suggesting a small increase in the conformational order of the peptoid. Cu(II) binding to both **7mer-HQ2** and **12mer-HQ4** produced new peaks between 240 nm and 280 nm, implying that chiral induction from the peptoid to the metal centers is attributed to the chirality of the oligomers and not to their helicity. The EPR spectra of both the Cu(II)–peptoid complexes suggested a tetrahedral distortion from the expected square planar coordination geometry, indicating that not only the chirality of the peptoid effects the Cu(II) center by establishing an asymmetric environment about it, but also the coordination geometry of the Cu(II) center is influenced by the peptoid backbone. Overall, it was concluded that metal-binding interactions alone could not lead to peptoid folding, suggesting that the presence of at least some chiral-bulky side chains might be also required. The results also suggested that more than one MB site is required to initiate peptoids folding.[30]

Therefore, in follow-up studies, the goal was to generate peptoid helices upon Cu(II) binding, utilizing the above conclusions. To do that, a set of Cu(II)-binding peptoids having much less than two thirds of

(a)

(b)

(c)

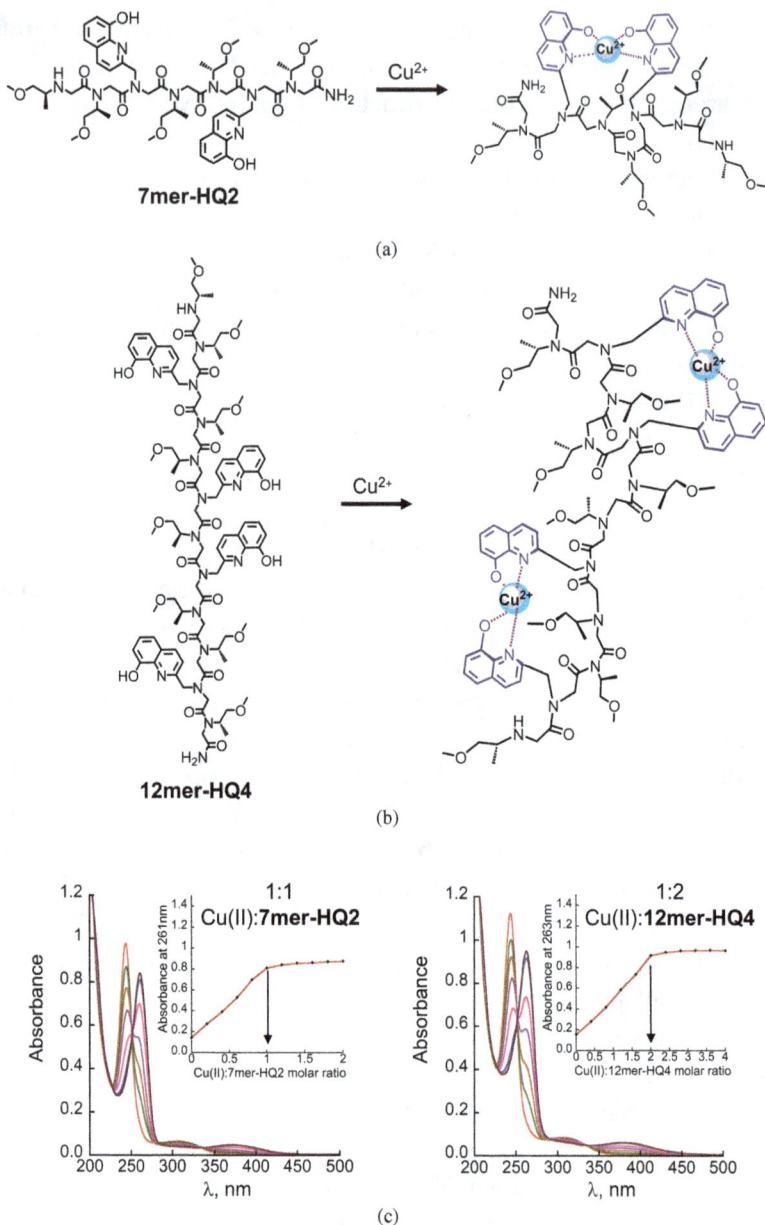

Figure 6. Schematic illustration of the peptoids **7mer-HQ2** (a) and **12mer-HQ4** and (b) forming intramolecular complex (**7mer-HQ2**)Cu and (**12mer-HQ4**)Cu$_2$ upon binding to Cu ions. (c) UV–vis titrations of **7mer-HQ2** and **12mer-HQ4** with Cu(II) ions. Insets: molar-to-ratio plots, suggesting the stoichiometric peptoid:Cu(II) ratio.

chiral-bulky substituents were designed based on the sequences of **7mer-HQ2** and **12mer-HQ4**. First, based the peptoid heptamer **7P1**, which was identical to **7mer-HQ2** except one Nsmp group at the 7th position that was replaced by one Nspe group. This modification was expected to slightly increase the conformational order of the peptoid and facilitate folding upon Cu(II) binding. Indeed, the CD spectra of **7P1** revealed low-intensity double minima, consistent with an initiation of a helix formation. Upon Cu(II) binding, the intensity of this signal decreased, and this decrease was attributed to structural changes within the peptoid, enforced by the coordination geometry of the Cu(II)–peptoid complex (Figure 7).[45]

Therefore, additional modifications of Nsmp groups by Nspe side chains were made, generating the set of peptoids **7P2–7P5** by keeping Nspe at 7th position constant and replacing and additional Nsmp group by an Nspe side chain in different positions along the peptoid scaffold. In addition, we designed peptoid **7P6**, having the two Nspe groups at the 3rd and 4th positions (between the two HQ ligands), forming a "bulky core." All peptoids were shown to bind Cu(II) in an intramolecular fashion, forming complexes of the type PCu, as revealed by UV–vis and ESI-MS. Systematic studies of the Cu–peptoid complexes by CD and EPR spectroscopies revealed the following clear trend: the larger the angle between the two HQ groups is (detected by the lower-exciton-coupled CD intensity), the larger is the distortion of the metal center from square planar geometry toward tetrahedral geometry (from EPR data) and the more ordered the peptoid secondary structure is (as seen by near-UV

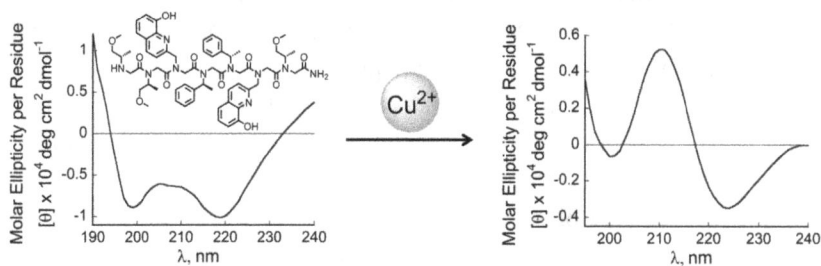

Figure 7.   Schematic illustration of peptoid **7P6** forming a PP-I-like helix upon binding to Cu(II) ions. Changes in the CD spectra of free peptoid **7P6** (left) and (**7P6**)Cu (right).

signals in the CD spectra). Surprisingly, coordination of Cu(II) to **7P6** resulted in a CD signal completely different from that of the free peptoid, exhibiting minima at 200 and 226 nm and maximum at 212 nm (Figure 7). Interestingly, this different spectrum resembles the spectrum of a PP-I peptide helix, similarly to previously reported CD spectra of the all-cis Ns1npe- and Ns1tbe-based PP-I peptoid helices.[17,21,24] We assumed that the driving force for the formation of this exceptionally ordered (**7P6**)Cu helix is the initial secondary structure of **7P6** as implied from the low-intensity double minima observed in its CD spectra combined with the coordination geometry of its Cu complex. Our findings reveal that the effect of Cu(II) binding on the peptoid structure depends both on the peptoid sequence (leading to the initial secondary structure), and on the preferred coordination geometry of the Cu(II) ion.

Following the two studies described above, the sequence of **12mer-HQ4** was modified to include two Nspe side chains instead of two Nsmp side chains and a new set peptoid dodecamers was designed. This new set, namely peptoids **12P1**, **12P2**, and **12P3**, had Nspe groups in the 1st and 12th positions, 2nd and 12th position, or 1st and 2nd positions, respectively, four HQ ligands in the 3rd and 6th positions and in the 8th and 11th positions forming two distinct binding sites, and six Nsmp groups at other positions along peptoid scaffold.[32] Based on rigorous spectroscopic measurements performed after the addition of Cu(II) to **12P1–12P3**, all three peptoids were shown to form diCu–peptoid complexes of the type PCu2 with Cu(II) metal center coordination geometry been pseudo-tetrahedral, in accordance with previously reported results for HQ-bearing peptoids. Importantly, results obtained from CD spectroscopy, solution NMR techniques and high-level DFT calculations demonstrated that **12P1** and **12P2**, which were shown to be unstructured peptoids, can fold upon Cu(II) binding to form helical structures. The structure of (**12P1**)Cu$_2$ is depicted in Figure 8. These results represent the first demonstration of generating a peptoid helix by metal coordination. Moreover, DFT-based calculated spectra and structures of metallopeptoids were obtained in this study for the first time. Interestingly, the peptoid **12P3**, which was also shown to be unstructured, did not fold upon Cu(II) binding. It was therefore concluded that the Nspe group at the C-terminal of peptoid dodecamers plays a crucial role in the stabilization of the helical secondary structure.

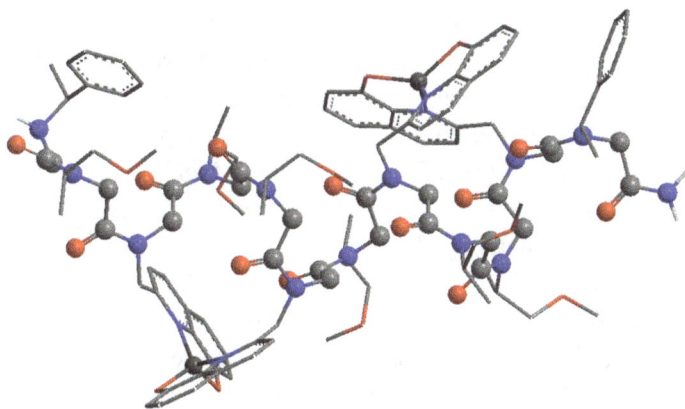

Figure 8. Side view of (**12P1**)Zn$_2$ optimized at the TPSS-D3/def2-TZVP + COSMO (methanol) level, assuming the right-handed cis PP-I helical structure. The main-chain C, N, O atoms and acidic H atoms (HQ hydroxyl and amide NH groups) are highlighted as grey, blue, red, and white balls, respectively.

B. Cu(II)-Mediated Peptoids Self-Assembly

In the previous section we have described our attempts to constrain the conformational order of peptoids by stabilizing their secondary (helical) structure in solution *via* their coordination to Cu(II) ions. Another known effective strategy for establishing conformational order of peptoids is macrocyclization.[52,53] The first peptoid macrocycles were formed *via* covalent bonds between pendent and/or backbone groups within the peptoid.[52,53] Notably, in nature, formation of high-order structures is often achieved *via* self-assembly rather than covalent interactions.[46] Moreover, metal–ligand coordination is a key interaction in the self-assembly of both biopolymers and synthetic oligomers.[46] Thus, the next step toward the development of highly constrained three-dimensional architectures from peptoids was to demonstrate that two or more peptoid segments can self-assemble *via* metal coordination to form high-ordered structures such as macrocycles, helical rods, nanotubes, and so on.

The first example of Cu(II)-mediated self-assembly of peptoids described a series of three self-assembled Cu(II)-peptoids and their unique crystal structures: highly symmetric macrocycles composed of two peptoid trimers held together *via* coordination to two Cu^{2+} ions, which two of these macrocycles are bridged by a water molecule (Figure 9).[46] The first macrocyclic duplex was assembled from two peptoid sequences incorporating Bipy as a ligand for Cu(II) binding, an ethanol group as an additional ligand, and non-coordinating benzyl group for structure directing and further stabilization of the macrocycle (Figure 9a). In attempts to gain control over the self-assembly process we further modified the ethanol group by a methoxy (Figure 9b) or an amine group (Figure 9c). After treating the peptoids with Cu(II) in methanol, the Cu–peptoid complexes were formed and crystallized. Single-crystal X-ray analysis showed that all three metallopeptoids formed Cu–peptoid duplexes of the type P_2Cu_2, while self-assembling into macrocyclic arrangements. Interestingly, while the ethanol group participated in the coordination of Cu(II) (Figure 9d), the methoxy and amine groups did not, and instead, the two Cu(II) ions within their corresponding Cu–peptoid complexes, bound one water molecule as an oxo-bridge between them, enabling

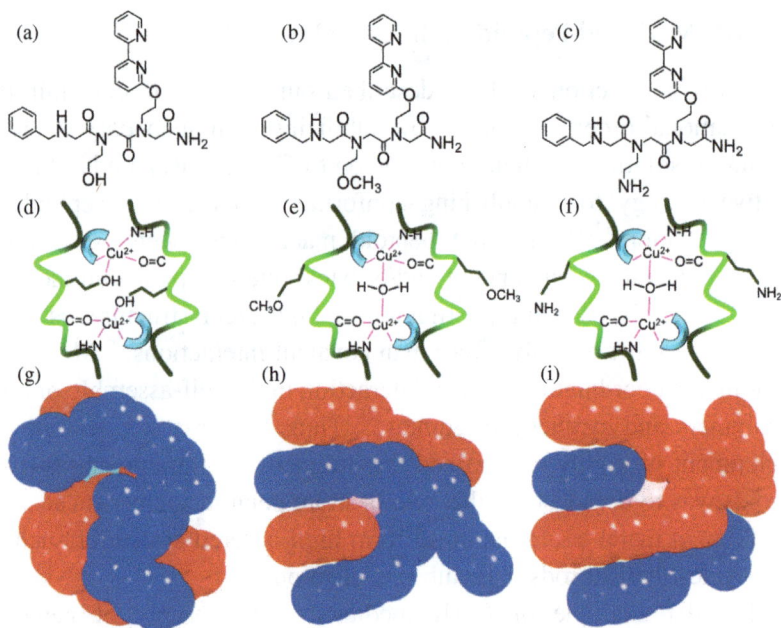

Figure 9. Peptoid sequences and representation of their corresponding complexes with Cu(II). Each color indicates one binding peptoid.

aromatic ring stacking (Figure 9e, f), which leads to a unique assembly resulting in the first metallopeptoid helicates (Figure 9h–i).[46] Notably, both experimental spectroscopic data and DFT calculations suggested that the Cu–peptoid macrocycles assembled from the peptoids having ethanol or methoxy groups are not present in solution but rather exist as monomeric complexes of the type PCu in acetonitrile, presenting a solid/solution state equilibrium. In contrast, the macrocycle assembled from the peptoid having the amine group does exist in solution.[46]

Although the macrocycles described above represent the first example of metal ions–mediated self-assembly of peptoids, they can still be defined as distinct structures as they did not pack nor stack to form more complex (and thus more biomimetic) supramolecular architectures. Indeed, targeted/controlled self-assembly of subunits toward the formation of distinct/desired supramolecular architectures is one of the most significant phenomena that happens in natural system and thus is highly significant in chemistry and biology.

In natural processes, peptide-based secondary structures form versatile turns, different helices, or multiple types of sheets by directed association of the peptide to produce 3D well-defined frameworks that work as membranes and/or channels.[54] These supramolecular architectures play significant role in several biochemical processes spanning recognition and/or transportation of biologically important ions and molecules.[55] Although there are several examples showing self-assembly of artificial peptides upon metal binding,[56–58] controlling this process toward a specific architecture is still challenging. Therefore, a worthy goal is controlling/directing the Cu(II)-mediated self-assembly of peptoids toward the controlled formation of various architectures.

To this aim, a set of peptoid trimers having Bipy in the second position within the sequence together with a pyridine (Pam) group in the N-terminal that acts as coordinating group, and a non-coordinating group at the C-terminus, were designed (Figure 10).[47] The Pam group in the N-terminal was included as it has the potential to interact with a benzyl or Pam group from another P_2Cu_2 unit via π-π interactions, such that it might enable further self-assembly to form supramolecular structures. For the monomer in C-terminus, various aromatic and aliphatic groups were

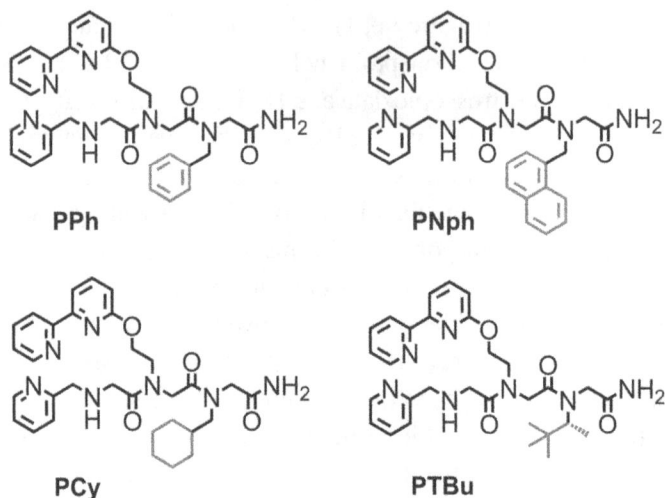

Figure 10. Peptoid sequences with varying substitution in C-terminus with Bipy in the second and pyridine in the third substitution.

chosen in order to evaluate the effect of this group on the self-assembly process, such that by controlling the type of this group (and thus the peptoid sequence) control over the self-assembled supramolecular architecture will also be achieved. The hypothesis was that while intermolecular interactions between an aromatic group and Pam might lead to one type of architecture, such interactions will not be possible with aliphatic groups, and as a result, different supramolecular structures might be formed in these cases.

Accordingly, a set of four peptoids was prepared (Figure 10), of which two have aromatic side chains, benzyl and naphthyl (**PPh** and **PNph**, respectively), and two have aliphatic side chains, cyclohexyl and tertiary butyl groups (**PCy** and **PTBu**, respectively) in the C-terminus. The peptoids were treated with Cu(II), resulting in Cu(II)–peptoid duplexes as evident by X-ray analysis. Interestingly, these were stable in solution as evident by ESI-MS study.

The crystals obtained for Cu(II)-**PPh/PNph** yielded a helical rod-like architecture as evident from the X-ray analysis. For Cu(II)-**PPh,** one distinct metallopeptoid duplex is shown in Figure 11a, which was further packed with other such duplexes *via* intermolecular π-π interactions

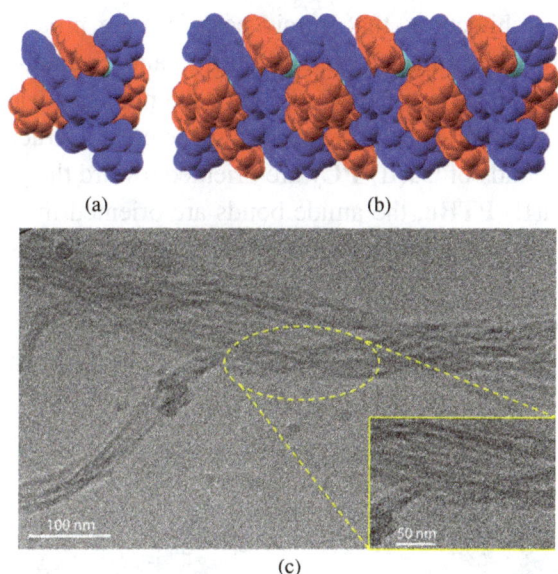

Figure 11. (a) Folded single metallopeptoid duplex of Cu^{2+} with peptoid having benzyl amine substitution. (b) Supramolecular rod-like helical strand formed by self-assembly of several metallopeptoid duplexes. (c) Cryo-TEM analysis of a sample crystal of Cu^{2+} with peptoid having benzyl amine substitution (100 μM).

between the non-coordinating benzyl moiety from one metallopeptoid and the pyridine group from adjacent unit, eventually forming the obtained supramolecular helical structure. The stability of metallopeptoid duplexes in solution prompts us to investigate the stability of the supramolecular architecture in solution. Therefore, for further insight regarding the solution state stability of the architectures, cryo-TEM analysis was performed with **Cu^{2+}-PPh**. Interestingly, similar helical structure as observed from X-ray study was also observed from cryo-TEM analysis. Helical segments were shown to aggregate into entangled fibrils and fold into bundles (Figure 11c). The width of each fiber was found approximately 11 nm that indicates stacking of several helices together. Crystallographic analysis of Cu(II)-**PNph** shows identical packing and supramolecular architecture as of Cu(II)-**PPh**.

Interestingly, in the cases of the Cu(II)–peptoids having an aliphatic substitution (**PCy** and **PTBu**, Figure 10), the obtained crystal structures demonstrated architectures that mimic pore-forming protein framework (Figure 12a, b). These metallopeptoids form nano channel–based structure with varying pore size. For Cu(II)-**PCy** the pore size was larger than Cu(II)-**PTBu** as obtained from X-ray study. The crystal structure revealed that all amide bonds of Cu(II)-**PCy** are oriented toward the pore, while in the case of Cu(II)-**PTBu**, the amide bonds are oriented in an alternating ("zigzag") fashion leading to a much tighter packing that eventually reduces the pore size. Cu(II)-**PCy** was chosen and analyzed by cryo-TEM and the monographs showed white spheres (circled in yellow, Figure 12c) with similar sizes to that of the pores obtained from the crystal structure.

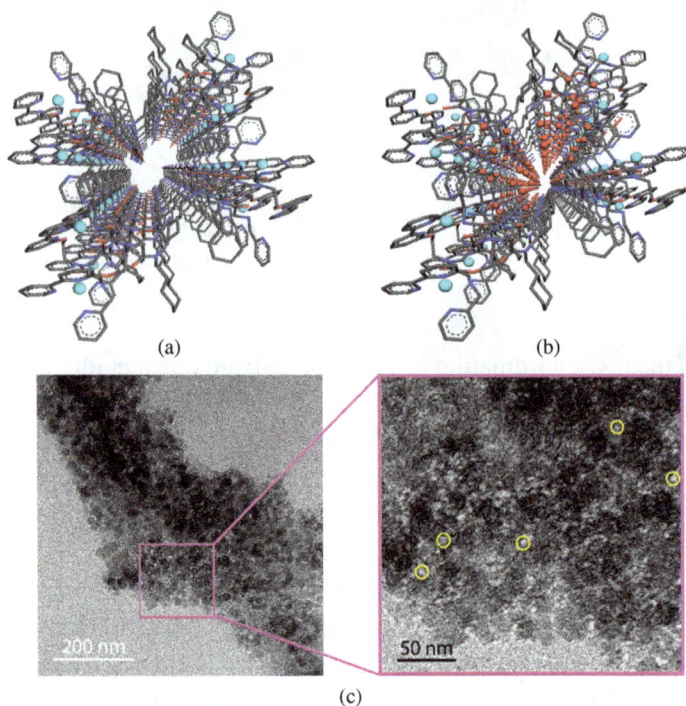

(a)   (b)

(c)

Figure 12. Spatial orientation of Cu(II) with peptoid having cyclohexyl amine substitution supramolecular cavity along the crystallographic axis in the absence (a) and presence (b) of water (Color code — red: oxygen; blue: nitrogen; cyan: Cu(II); and grey: carbon. Hydrogen atoms and perchlorate ions are removed for clarity). (c) Cryo-TEM of the crystals (100 μM in water) with expanded view marked framed in purple.

Thus these were suggested to be the channel pores. Overall, for the first time, controlling the sequence by simply modifying the non-coordinating group at the C-terminus lead to control over the supramolecular architecture of metallopeptoids.

IV. From Structure to Function: Cu-Binding Peptoids and Cu–Peptoids as Functional Materials

A. Selective Recognition of Small Molecules by Self-Assembled Cu(II)–Peptoids

The Cu–peptoid nano-channels described in section III.B (Figure 12), which were self-assembled from the peptoids **PCy** and **PTBu** (Figure 10) were found to be stable in solution as evident from the cryo-TEM and ITC experiments. This allowed further exploration of their (selective) recognition properties toward biologically relevant small molecules and anions. The recognition and transport of ions and small molecules by biological channels are important processes for regulation of chemicals within the cells. Mimicking such processes is therefore a key to the development of specific receptors for therapeutic applications.[59,60] Selective recognition of ions and small biomolecules requires control over the pore size of the nano-channels and their ability to recognize guest ions and small molecules by non-covalent interactions. As amide groups can interact with ions and small biomolecules, the host–guest interactions and recognition ability of the two types of nano-channels, formed by either Cu(II)-**PCy** and Cu(II)-**PTBu** toward the anions Cl^-, $H_2PO_4^-$, CO_3^{2-}, PO_4^{3-}, citrate, and glutamate, and the neutral small biomolecules glucose, glycerol, arabinose, and mannose in solution were explored. The host–guest interactions between Cu–peptoid nano-channels and anions were initially monitored by following the changes in the vibrational stretching of the amide bond ($\nu = {\sim}1600$ cm^{-1}) using IR spectroscopy. The results from these experiments suggested that there is an interaction, and possible binding, between Cu(II)-**PCy** and the smaller anions, while the larger anions citrate and glutamate did not interact with the metallopeptoid host. Interestingly, similar measurements using Cu(II)-**PTBu** resulted in the expected shift only upon the addition of Cl^-, while no change was detected upon the addition of $H_2PO_4^-$. These results indicated that nano-channels assembled from Cu(II)-**PCy**, having larger pores are a non-selective host for Cl^-, whereas the nano-channels assembled from Cu(II)-**PTBu** interact with Cl^- only.

The host–guest interactions between Cu–peptoid nano-channels and anions were initially monitored by gradually increasing the concentration of glycerol in the solution containing Cu(II)-**PCy** resulted in a gradual shift of the vibrational stretching near 1605 cm^{-1} to a higher wavenumber near 1616 cm^{-1} while gradual addition of glucose, arabinose, or mannose did not lead to any shift in the FT-IR spectra. These results suggested that there is a selective interaction, and possible selective binding, between Cu(II)-**PCy** and glycerol, while the larger, sugar molecules, do not inter-act with the metallopeptoid host. Similar titrations to the solution of Cu(II)-**PTBu** did not lead to changes in the vibrational stretching of the metallopeptoids, suggesting that the pore size of the corresponding nano-channels is too small to host such biomolecules.

The interaction between the Cu–peptoid nano-channels and glycerol was further verified by ^{1}H-NMR study, in which Cu(II)-**PCy** or Cu(II)-**PTBu** were added to D_2O solutions of either glycerol or glucose. The ^{1}H-NMR spectra of glycerol and glucose were measured and monitored before and after the addition of the Cu–peptoids. The spectra showed that while there was no change in the chemical shift of glucose upon the addi-tion of each Cu–peptoid solution or in the chemical shift of glycerol after the addition of Cu(II)-**PTBu** (Figure 13a, b), a clear downfield shift was observed after the addition of Cu(II)-**PCy** to glycerol (Figure 13c, d), sup-porting the results obtained from the FT-IR analysis. Furthermore, the sample solution containing the Cu(II)-**PCy** loaded with glycerol was lyo-philized and the microcrystalline solid obtained was washed thoroughly with methanol to remove residues of unbound dissolved glycerol (if any), and dried in vacuum. The dry solid was solubilized in diluted hydrochloric acid and treated with bicarbonate, and a sample from this solution was analyzed by ESI-MS. The ESI-MS analysis showed a signal indicating the mass of glycerol, indicating again the interaction of Cu(II)-**PCy** with glycerol. Similar workup and analysis performed on solution sample of Cu(II)-**PTBu** loaded with glucose did not reveal any trace of glucose in

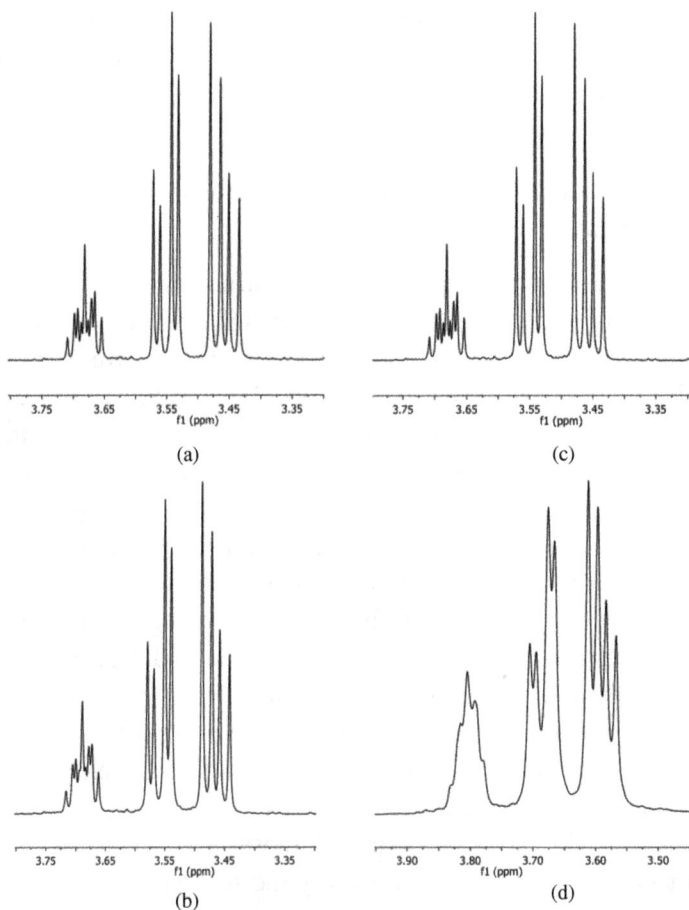

Figure 13. ^{1}H-NMR of glycerol in D_2O (a) before and (b) after addition of Cu(II)-**PTBu** and (c) before and (d) after addition of Cu(II)-**PCy** (host and guest concentration is 10 mM).

the ESI-MS spectrum, supporting the results obtained from the FT-IR and ^{1}H-NMR experiments indicating no glucose recognition.

Overall, these results demonstrate that the self-assembled nano-channels can interact with biologically relevant anions or molecules and that by controlling their pore size we can regulate interactions with specific anions and small biomolecules toward the use of these channels as selective receptors. The selective binding of the anions and small biomolecules to Cu(II)-**PCy** and Cu(II)-**PTBu** also validate the self-assembly of the Cu–peptoids in solution.

B. Selective Recognition of Cu(II) by Peptoids Toward Drug Design

Following the understanding regarding the structural aspects of Cu(II)–peptoids and the sequence requirements for high affinity to Cu(II), selective chelators to Cu(II) could be designed toward selective extraction of Cu(II) from biomimetic and biological media, as well as from Cu–proteins. As Cu is an essential trace element, involved as a cofactor in many proteins, its concentration level is typically tightly regulated within cells. However, Cu could be also toxic due to its redox ability, and its misregulation is usually associated with two rare genetic disorders: Wilson's disease (copper overload) and Menkes disease (copper deficiency).[61,62] Moreover, copper was found to play a role in several neurodegenerative diseases, including Alzheimer's and Parkinson's diseases, and high levels of Cu ions are usually found in serum and tissue samples of cancer tumors. One of the therapeutic approaches to restore balance of copper is to extract the excess copper by chelation, that is, molecules with high selectivity to Cu ions. The field of copper chelation accounts for hundreds of known suitable molecules,[63,64] however, many of those still do not meet all the requirements for carrying out efficiently the recognition process. Thus, there is still a demand for the new ligands to be developed. One of the recent approaches is the development of peptidomimetics as selective chelators for Cu, as they combine the advantages of currently known ligands with improved bioavailability properties. Among peptidomimetics, peptoids are specifically interesting due to their excellent pharmacokinetic properties, such as long half-lives and high membrane permeability.

It was previously shown, already in the reported first example of Cu(II)–peptoids,[15] that the pre-organization of two HQ ligands at the *i* and *i* + 3 positions of a helical (Nspe-based) peptoid, lead to a very large

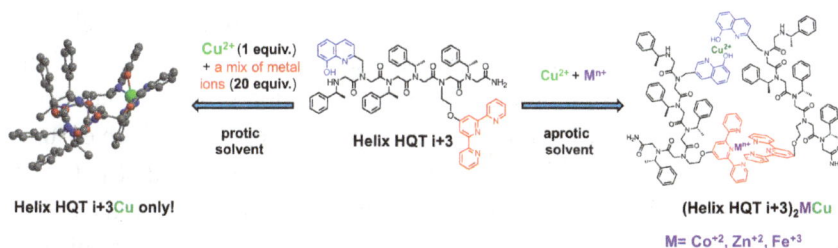

Figure 14. A schematic summary of Cu(II) binding to the helical peptoid **Helix i + 3**, which is capable of intramolecular binding of Cu(II) ions or intermolecular binding of two different metal ions in a selective manner.

binding constant observed for the Cu(II)–peptoid complex ($\log K = 14$, in MeOH:H$_2$O (4:1) solution[15]). This observation established the requirement for intramolecular binding of the Cu(II) by two MB ligands that are pre-organized at the same site of the peptoid's helix in order to achieve exceptionally high affinity to Cu(II). Capitalizing on this requirement, the helical Nspe-peptoid hexamer **Helix i + 3**, bearing the MB ligands HQ and Terpy at the i and $i + 3$ positions (Figure 14), was designed and synthesized.[31] Upon Cu(II) binding, the (**Helix i + 3**)Cu(II) was generated, and its spectroscopic analysis suggested a pentagonal binding environment about the Cu(II) and high association constant was evaluated ($\log K = 13$, in MeOH:H$_2$O (4:1) solution). Moreover, in methanol, **Helix i + 3** was shown to bind 1 equiv. of Cu(II), selectively, from a mixture of excess up to 20 equiv. of other metal ions, as confirmed by UV–vis, ICP-MS, and ESI-MS studies. The unique selectivity for Cu(II) was shown to be a result of the combination of both the choice of the two specific MB ligands and their pre-organization within the peptoid helix. Interestingly, in aprotic solvents, such as acetonitrile, a heterobimetallic peptoid duplex of the type P$_2$CuM was obtained when **Helix i + 3** was treated with a solution mixture of Cu(II) and another metal ion such as Zn(II) and Co(II) (Figure 14).

Unfortunately, the Nspe side chains are hydrophobic, thus limiting the solubility of the peptoid in water due to its high overall hydrophobicity.

As a result, possible utilization of **Helix i + 3**, with its remarkable selectivity to Cu(II), as a drug candidate, required further optimization of the sequence. A possible solution was the incorporation of one or more piperazine units within the backbone of hydrophobic peptoids, which ensures their water solubility, while maintaining their sequence and structure.[65] This should initiate opportunities for future biochemical studies with modified **Helix i + 3** toward selective chelation in the context of disease therapeutics.

Another possibility for the design of water-soluble-selective peptoid chelators was to use short peptoids with a different set of MB ligands. Thus, the water-soluble peptoid trimer **PCA-Nspe**, having an Nspe side chain at its C-terminal, followed by Bipy) and Npam (at the N-terminus) as a MB ligands was recently synthesized and evaluated for metal coordination.[48] **PCA-Nspe** was shown to selectively bind Cu(II) in the presence of high excess of 40 equiv. of other metal ions, forming an intramolecular complex of the type PCu, as confirmed by UV–vis, CD, EPR, and ESI-MS analysis. Moreover, **PCA-Nspe** could extract Cu(II) from the natural copper-binding protein metallothionein, as indicated from the CD spectroscopy (supported by LC-MS analysis) before and after the addition of **PCA-Nspe**. Computational studies suggested the structure-directing Nspe group plays the crucial role in pre-organization of the MB ligand in **PCA-Nspe**, which results in high selectivity. Interestingly, replacing the Nspe side chain with a non-chiral benzyl amine side chain, resulted in an intermolecular binding of Cu(II), forming a complex of the type P_2Cu_2. This intriguing observation prompted further sequence–function relationship studies aiming to understand the role of the non-coordinating side chain in the selectivity to Cu(II).[49] Consequently, a set of peptoids, varied in their side chain at the C-terminal, using different aromatic and aliphatic chiral and achiral side-chains, was synthesized and its metal coordination was estimated and investigated. The results suggested that the substitution at the C-terminal has a significant effect on the selectivity to Cu(II), and that high selectivity requires not only chirality and bulkiness, but rather the combination of both, together with a specific position of the non-coordinating side chain along the peptoid sequence. Moreover, one of these chelators, **PC1**, having an Ns1npe side chain instead of Nspe, could

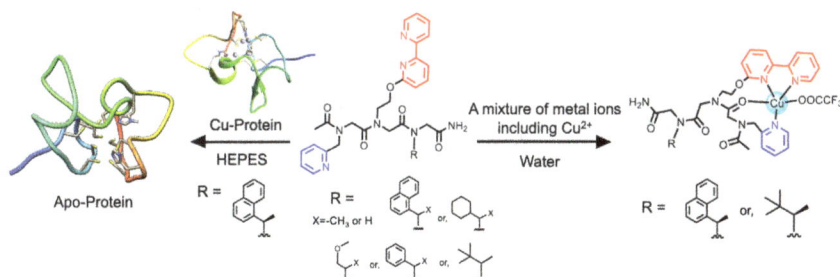

Figure 15. A cartoon representation of Cu(II) chelation by PC peptoids from a copper-containing protein (left) or selective recognition from the mixture of other metal ions (right).

extract Cu(II) from metallothionein, as indicated from CD spectroscopy (Figure 15).

C. Positive Allosteric Cooperativity in Metal Binding to Cu(II)–Peptoid

The high selectivity of **Helix i + 3** to Cu(II) (see section III.A), prompted the idea to replace one of the HQ ligands in the peptoid **12P1** with a Terpy ligand, forming a peptoid with two distinct binding sites. This was anticipated to enable selectivity to Cu(II) in the HQ/Terpy site leaving the HQ/HQ site available for the binding of a different metal ion, resulting in a peptoid complex of the type PCuM. Consequently, the peptoids **12P5** and **12P6** (Figure 16a) were synthesized, characterized, and treated with Cu(II). CD spectroscopy of **12P5** revealed that coordination of Cu(II) to the Terpy/HQ binding site initiated folding, followed by the pre-organization of the HQ/HQ binding site, and by this facilitated the coordination of Zn(II) or Co(II) ions to the HQ/HQ site, demonstrating positive allosteric cooperativity in the binding of Zn(II) or Co(II) (Figure 16b).[32] In contrast, the unstructured **12P6** did not fold upon Cu(II) binding and the corresponding Cu–peptoid complex did not facilitate the binding of a second metal ion (Figure 16c). Overall, this study demonstrated for the first time that a structural change in a peptoid, induced by the selective metal binding of Cu(II), leads to positive allostery.

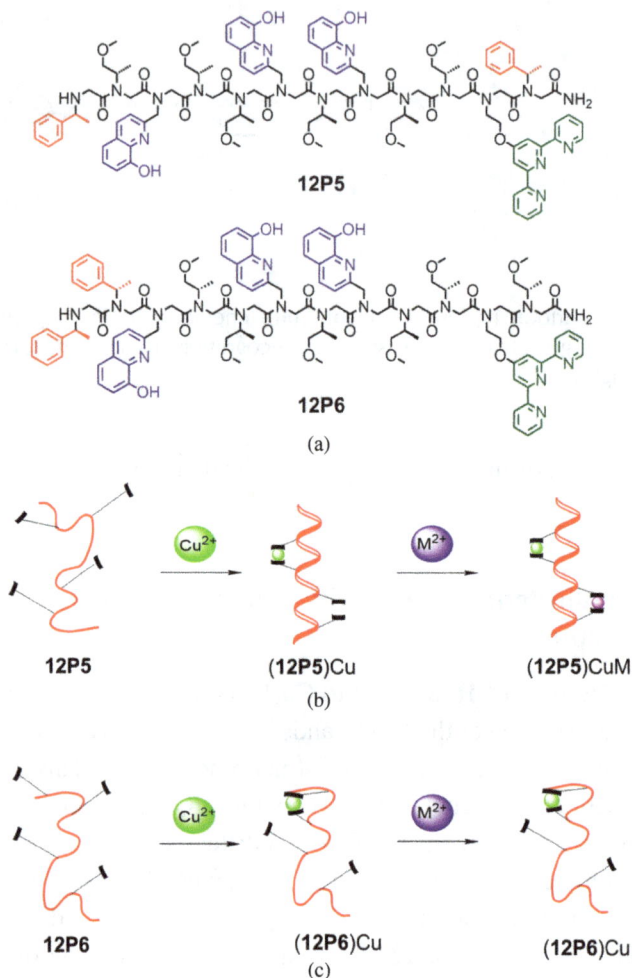

Figure 16. (a) Peptoid sequences of 12mer peptoids **12P5** and **12P6**. (b) A cartoon demonstrating positive allosteric cooperative binding to (**12P5**)Cu, compared to (c) no allosteric cooperativity of (**12P6**)Cu.

D. Cu(II)–Peptoid Complexes as Versatile, Efficient, and Selective Bio-Inspired Catalysts

Enzymatic catalysis is one of the most important functions carried out by natural metalloproteins. The catalytic activity of enzymes is largely dependent on the cooperativity between binding sites, creating catalytic pockets, which enable enzyme specificity and efficiency. A promising approach for the design of synthetic catalytic systems that can imitate enzymatic efficiency is to mimic the intramolecular cooperativity of enzymes and the secondary coordination sphere about the catalytic metal site, which facilitate its activity. This can be potentially done by designing artificial peptidomimetic catalytic systems, containing functional groups responsible for catalytic activity, as well as structural elements, in close proximity to each other within one oligomer. This should create a confined catalytic pocket together with its second coordination sphere, similar to the ones observed in nature.

The first attempt to mimic such enzymatic cooperativity between two catalytic groups within a peptoid, capitalized on the oxidation of alcohols to aldehydes by a Cu-Bipy catalysts that had utilized (2,2,6,6-tetramethyl-piperidin-1-yl)oxyl (TEMPO) as a co-catalyst, N-methyl imidazole as a nucleophile, and molecular oxygen as an oxidizer (Figure 17a, left).[66] Based on this intermolecular cooperative catalytic system, a set of peptoid trimers incorporating 1,10-phenanthroline as a ligand for Cu at the N-terminus, TEMPO in the respective position, and a non-catalytic aromatic or alkyl group at the C-terminus, as intramolecular cooperative catalytic systems for aerobic oxidation of alcohols, in the same reaction conditions as the intermolecular cooperative catalytic system but with lower catalyst loading.[67] All the peptoids showed catalytic activity, which

Figure 17. Oxidation of alcohols to aldehydes and oxidative coupling of alcohols and amines to imines catalyzed by an intermolecular cooperative catalytic system (a) or by the peptoid-based intramolecular cooperative catalytic system (b).

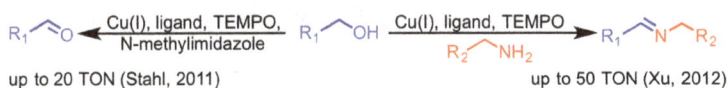

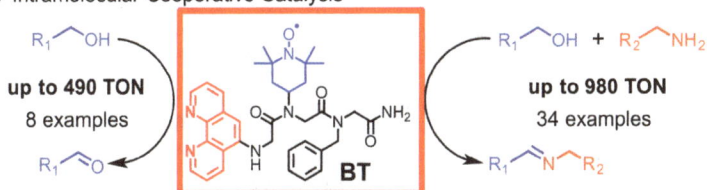

was at least 10 times higher than the intermolecular cooperative catalytic system or a control peptoid dimer (**DI**) that did not have the non-catalytic group. Moreover, the position of the three groups relative to each other also had a role in the catalytic activity. These results demonstrated that the pre-organization of the catalytic groups within the peptoid, and the presence of a non-catalytic group have a crucial role in the activity, suggesting that these enable the intramolecular mode of action similar to the intramolecular cooperativity in enzymes. The trimer **BT**, having a benzyl group at the C-terminus, combined with Cu, could catalyze the oxidation of various benzylic, allylic, and aliphatic primary alcohols with a TON of up to 16 times higher than a mixture of the two catalytic groups or the peptoid dimer that is lacking the non-catalytic group (Figure 17b, left).

Following studies revealed that the catalyst CuBT could also catalyze the oxidative coupling of alcohols and amines for the production of benzyl, aryl, heteroaryl, allylic, and aliphatic imines[68] with a TON up to 45 times higher than this achieved when phenanthroline, Cu, and TEMPO are mixed in solution[69] (Figure 17a, b, right). Recently, in a more subsequent research, resin-bound **BT** and **DI** (**TG-BT** and **TG-DI**) were presented as the first insoluble and recyclable catalytic peptoids. It was shown that although **TG-BT** also operates *via* the intra-peptoid cooperativity mode, it is less efficient than **BT**, even after several cycles. On the other hand, **TG-DI** is far more active than **DI** (than **BT**) and can be recycled several times to enable a much higher turnover number. These studies revealed that **TG-DI** operates *via* an intra-resin cooperativity mode, which enables its high activity compared with **DI**, **BT**, and **TG-BT** (Figure 18).[70]

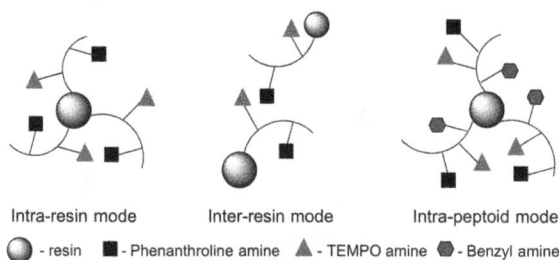

Intra-resin mode   Inter-resin mode   Intra-peptoid mode

⬤ - resin   ◼ - Phenanthroline amine   ▲ - TEMPO amine   ⬤ - Benzyl amine

Figure 18. The three different cooperativity modes suggested for the catalytic activity of **TG-DI** (intra-resin and inter-resin) and of **TG-BT** (intra-peptoid).

An important task in the field of catalysis, which is also a very challenging one, is the development of efficient electro- and photocatalysts for splitting water to molecular oxygen and hydrogen, the latter being a renewable and sustainable energy source. The first step of this catalytic reaction, water oxidation ($2H_2O \rightarrow O_2 + 4H^+ + 4e^-$; $\Delta E° = 1.23$ V, $\Delta G = 113$ kcal mol^{-1}), is the thermodynamically challenging step, both in natural and especially in artificial systems, due to the transfer of four protons and four electrons in a relatively low potential (0.876 V at pH 6).[71] In nature, this transformation is catalyzed by the oxygen-evolving complex (OEC) in photosystem II of green plants and cyanobacteria, utilizing solar energy and a manganese-oxo complex that its oxidation toward water oxidation is facilitated by amino acids such as tyrosine and histidine from the second coordination sphere.[72]

Based on the catalytic capabilities of **BT** and of the Cu-Bipy complexes that act as efficient electrocatalysts for water oxidation, albeit at high pH, with high overpotential and most importantly which were unstable under the electrocatalytic conditions,[72,73] a Cu–peptoid was recently designed and reported as the first peptoid-based electrocatalyst for water oxidation.[13] This peptoid trimer (**BPT**) is composed of a Bipy group as a

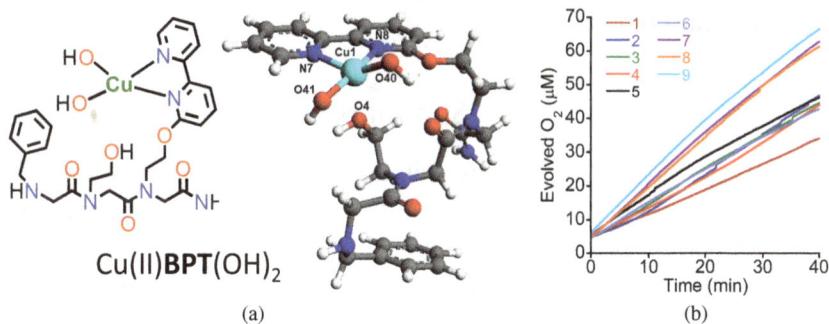

Figure 19. (a) Molecular structure of Cu(II)**BPT**(OH)$_2$ and its geometry-optimized structure (optimized by DFT-D3 calculations (considering the dispersion correction) at the level of B3-LYP with def2-TZVP for Cu(II) and def2-SVP for the other atoms as basis sets with unrestricted Hartree–Fock, using turbomole and ORCA (3.0.3) software package. (b) Evolution of O$_2$ and charge accumulation during nine runs (40 min. each) in phosphate buffer solution (0.1 M) at pH 11.5 with 0.5 mM Cu(II) **BPT**(OH)$_2$. Oxygen was measured with a fluorescence probe. Porous glassy carbon electrode was used at 1.35 V *versus* NHE.

ligand for Cu(II), an –OH group, as a tyrosine mimic, and a benzyl group; the two non-catalytic groups were designed to act as a second coordination mimic aiming to stabilize the Cu(II)/Cu(III) center and facilitate its activity. Upon treating the peptoid with Cu(II) in alkaline buffer, the complex Cu(II)**BPT**(OH)$_2$ was formed and both experimental and computational data revealed that Cu(II) is bound to **BPT** *via* Bipy and two hydroxyl ions originating from the basic solution. At pH 11, the catalyst was stable over at least 15 hours of electrolysis and could be reused for at least nine times in 40-minute runs, resulting in an overall TON of ~56 within 6 hours, if the pH was readjusted after each run. Electrochemical experiments revealed that the reversible Cu$^{III/II}$ oxidation wave, which is a key for water oxidation, occurred at an unusually low $E_{1/2}$ of +0.30 V *versus* NHE in contrast to a peptoid analog, in which the ethanolic –OH group was replaced by –OCH$_3$ to give a potential of +0.50 V for the Cu$^{III/II}$ transition. Based on these as well as other electrochemical experiments, spectroscopic data, and DFT calculations, a stable key peroxide intermediate was identified and an intramolecular cooperative catalytic pathway was proposed, suggesting that the proximal –OH group and the etheric oxygen atom attached to the Bipy moiety form strong hydrogen bonding with the coordinated hydroxide groups, thus has a major role in the high stability of the complex.

V. Summary

The incorporation of MB ligands within peptoids, first demonstrated in 2009 by Maayan *et al.* has opened up the new field of metallopeptoids. Extensive research on Cu(II) coordination has expanded the field of peptoids and of peptidomimetics in general, beyond structure and structural requirements and considerations, toward function. The broad understanding regarding the role of Cu(II) in peptoid folding and self-assembly led to the development of Cu(II)–peptoids as functional materials that are inspired by natural Cu(II)-binding peptides and proteins. This includes the ability to mimic complicated biological activities such as selective recognition, allosteric cooperativity, and intramolecular cooperative catalysis similar to the one performed by enzymes. Although exciting, these achievements represent only the initial steps toward the many possibilities to utilize Cu for obtaining functional peptoids. Further advances in this field include studies on Cu(I) coordination to peptoids, utilization of Cu(I) and Cu(II)-binding peptoids as therapeutics, and the development of mono- and multi-Cu–peptoids as efficient and selective catalysts for various transformations, including ones that are currently too challenging to be performed by molecular catalysts.

VI. References

1. Zastrow, M. L.; Pecoraro, V. L. *Coord. Chem. Rev.* **2013**, *257*(17–18), 2565–2588.
2. Crichton, R. R. *Biological Inorganic Chemistry*, 2nd Ed., Ed. Crichton, R. R., Chapter 14 "Copper — Coping with Dioxygen." Elsevier, 2012, 279–296.
3. Bush, A. I. *Curr. Opin. Chem. Biol.* **2000**, *4*, 91–184.
4. Ala, A.; Walker, A. P.; Ashkan, K.; Dooley, J. S.; Schilsky, M. L. *Lancet* **2007**, *369*, 397–408.
5. Lovejoy, D. B.; Jansson, P. J.; Brunk, U. T.; Wong, J.; Ponka, P.; Richardson, D. R. *Cancer Res.* **2011**, *71*, 5871–5880.
6. Andersen, O. *Chem. Rev.* **1999**, *99*, 2683–2710.
7. Kalia, K.; Flora, S. J. S. *J. Occup. Health.* **2005**, *47*, 1–21.
8. Gellman, S. H. *Acc. Chem. Res.* **1998**, *31*(4), 173–180.
9. Huc, I. *Eur. J. Org. Chem.* **2004**, 17–29.
10. Hill, D. J.; Mio, M. J.; Prince, R. B.; Hughes, T. S. and Moore, J. S. *Chem. Rev.* **2001**, *101* (12), 3893–4012.
11. Seebach, D. and Gardiner, J. *Acc. Chem. Res.* **2008**, *41*(10), 1366–1375.
12. Cheng, R. P.; Gellman, S. H.; DeGrado, W. F. *Chem. Rev.* **2001**, *101*(10), 3219–3232.
13. Ghosh, T.; Ghosh, P.; Maayan, G., *ACS Catal.* **2018**, *8*(11), 10631–10640.
14. Mohan, D. C.; Ghosh, P.; Ghosh, T.; Maayan, G., *Chem. Eur. J.* **2020**, *26*, 9573–9579.
15. Maayan, G.; Ward, M. D.; Kirshenbaum, K. *Proc. Natl. Acad. Sci. U.S.A.* **2009**, *106* (33), 13679–13684.
16. Schettini, R.; De Riccardis, F.; Della Sala, G.; Izzo, I. *J. Org. Chem.* **2016**, *81*, 2494–2505.
17. Kirshenbaum, K.; Barron, A. E.; Goldsmith, R. A.; Armand, P.; Bradley E. K.; Truong, K. T. V.; Dill, K. A.; Cohen, F. E.; Zuckermann, R. N. *Proc. Natl. Acad. Sci. U.S.A.* **1998**, *95*(8), 4303–4308.
18. Ruan, G.; Engelberg, L.; Ghosh, P.; Maayan, G. *Chem. Commun.* **2021**, *57*, 939–942.
19. Miller, S. M.; Simon, R. J.; Ng, S.; Zuckermann, R. N.; Kerr, J. M.; Moos, W. H. *Drug Dev. Res.* **1995**, *35*, 20–32.
20. Kwon, Y.; Kodadek, T. *J. Am. Chem. Soc.* **2007**, *129*(6), 1508–1509.
21. Roy, O.; Dumonteil, G.; Faure, S.; Jouffret, L.; Kriznik, A.; Taillefumier, C. *J. Am. Chem. Soc.* **2017**, *139*(38), 13533–13540.

22. Zuckermann, R. N.; Kerr, J. M.; Kent, S. B. H.; Moos W. H. *J. Am. Chem. Soc.* **1992**, *114*(26), 10646–10647.

23. Culf, A. S.; Ouellette, R. J. *Molecules.* **2010**, *15*(8), 5282–5335.

24. Stringer, J. R.; Crapster, J. A.; Guzei, I. A.; Blackwell, H. E. *J. Am. Chem. Soc.* **2011**, *133*(39), 15559–15567.

25. Pokorski, J. K.; Miller Jenkins, L. M.; Feng, H.; Durell, S. R.; Bai, Y.; Appella, D. H. *Org. Lett.* **2007**, *9*(12), 2381–2383.

26. Wu, C. W.; Sanborn, T. J.; Zuckermann, R. N.; Barron A. E. *J. Am. Chem. Soc.* **2001**, *123*(13), 2958–2963.

27. Baskin, M.; Panz, L.; Maayan, G. *Chem. Commun.* **2016**, *52*, 10350–10353.

28. Wu, C. W.; Sanborn, T. J.; Huang, K.; Zuckermann, R. N.; Barron, A. E. *J. Am. Chem. Soc.* **2001**, *123*(28), 6778–6784.

29. Maayan, G.; Ward, M. D.; Kirshenbaum, K. *Chem. Commun.* **2009**, *2009*, 56–58.

30. Baskin, M.; Maayan, G. *Biopolymers.* **2015**, *104*, 577–584.

31. Baskin, M.; Maayan, G. *Chem. Sci.* **2016**, *7*, 2809–2820.

32. Baskin, M.; Zhu, H.; Qu, Z. W.; Chill, J. H.; Grimme, S.; Maayan, G. *Chem. Sci.* **2019**, *10*, 620–632.

33. Pearson, R. G. *J. Am. Chem. Soc.***1963**, *85*(22), 3533–3539.

34. Rubino, J. T.; Franz, K. J. *J. Inorg. Biochem.* **2012**, *107*, 129–143.

35. Griffith, J. S.; Orgel, L. E. *Q. Rev. Chem. Soc.* *11*, 381–393.

36. Pflaum, R. T.; Brandt, W. W. *J. Am. Chem. Soc.* **1954**, *76*(24), 6215–6219.

37. Osamu, Y.; Hiroshi, B.; Akitsugu, N. *Bull. Chem. Soc. Jpn.* **1973**, *46*(11), 3458–3462.

38. Johnston, W. D.; Freiser, H. *J. Am. Chem. Soc.* **1952**, *74*(21), 5239–5242.

39. Fleischel, O.; Wu, N.; Petitjean, A. *Chem. Commun.* **2010**, *46*, 8454–8456.

40. Galezowska, J.; Boratynski, P. J.; Kowalczyk, R.; Lipke, K.; and Czapor-Irzabek, H. *Polyhedron.* **2017**, *121*, 1–8.

41. Maayan, G.; Yoo, B.; Kirshenbaum, K. *Tetrahedron Lett.* **2008**, *49*(2), 335–338.

42. Zabrodski, T.; Baskin, M.; Kaniraj, P. J.; Maayan, G. *Synlett.* **2015**, *26*(04), 461–466.

43. Ford, B. K.; Hamza, M.; Rabenstein, D. L. *Biochemistry.* **2013**, *52*(21), 3773–3780.

44. Ghosh, P.; Maayan, G. *Chem. Sci.* **2020**, *11*, 10127–10134.

45. Zborovsky, L.; Smolyakova, A.; Baskin, M.; Maayan, G. *Chem. Eur. J.* **2018**, *24*, 1159–1167.

46. Ghosh, T.; Fridman, N.; Kosa, M.; Maayan, G. *Angew. Chem. Int. Ed.* **2018**, *57*(26), 7703–7708.

47. Ghosh, P.; Fridman, N.; Maayan, G. *Chem. Eur. J.* **2021**, *27*, 634–640.

48. Ghosh, P.; Maayan, G. *Chem. Eur. J.* **2021**, *27*, 1383–1389.

49. Ghosh, P.; Rozenberg, I.; Maayan, G. *J. Inorg. Biochem,* **2021**, *217*, 111388.

50. Sakaguchi, U.; Addison, A. W. *J. Chem. Soc. Dalton Trans.* **1979**, 600–608.

51. Garriba, E. and Micera, G. *J. Chem. Educ.* **2006**, *83*(8), 1229.

52. Macedi, E.; Meli, A.; De Riccardis, F.; Rossi, P.; Smith, V. J.; Barbour, L. J.; Izzo, I.; Tedesco, C. *Cryst. Eng. Comm.* **2017**, *19*, 4704–4708.

53. Shin, S. B. Y.; Yoo, B.; Todaro, L. J.; Kirshenbaum, K. *J. Am. Chem. Soc.* **2007**, *129*, 3218–3225.
54. Hu, C.; Chan, S. I.; Sawyer, E. B.; Yu, Y.; Wang, J. *Chem. Soc. Rev.* **2014**, *43*, 6498.
55. Li, J. *Supramolecular Chemistry of Biomimetic Systems.* Singapore, Springer, 2017.
56. Sawada, T.; Yamagami, M.; Ohara, K.; Yamaguchi, K.; Fujita, M. *Angew. Chem. Int. Ed.* **2016**, *55*, 4519.
57. Sawada, T.; Yamagami, M.; Akinaga, S.; Miyaji, T.; Fujita, M. *Chem. Asian J.* **2017**, *12*, 1715.
58. Kwon, S.; Shin, H. S.; Gong, J.; Eom, J. H.; Jeon, A.; Yoo, S. H.; Chung, I. S.; Cho, S. J.; Lee, H. S. *J. Am. Chem. Soc.* **2011**, *133*, 17618.
59. Moore, S. J.; Wenzel, M.; Light, M. E.; Morley, R.; Bradberry, S. J.; Gómez-Iglesias, P.; Soto-Cerrato, V.; Pérez-Tomas, R.; Gale, P. A., *Chem. Sci.* **2012**, *3*(8), 2501–2509.
60. Valkenier, H.; Judd, L. W.; Li, H.; Hussain, S.; Sheppard, D. N.; Davis, A. P. *J. Am. Chem. Soc.* **2014**, *136*(35), 12507–12512.
61. Hordyjewska, A.; Popiołek, Ł.; Kocot, J. *Bio. Metals.* **2014**, *27*, 611–621.
62. Lee, S.; Barin, G.; Ackerman, C. M.; Muchenditsi, A.; Xu, J.; Reimer, J. A.; Lutsenko, S.; Long, J. R.; Chang, C. J. *J. Am. Chem. Soc.* **2016**, *138*, 7603–7609.
63. Savelieff, M. G.; Nam, G.; Kang, J.; Lee, H. J.; Lee, M.; Lim, M. H. *Chem. Rev.* **2019**, *119*(2), 1221–1322.
64. Esmieu, C.; Guettas, D.; Conte-daban, A.; Sabater, L.; Faller, P.; Hureau, C. *Inorg. Chem.* **2019**, *58*(20), 13509–13527.
65. Mohan, C. D.; Kaniraj, P. J.; Maayan, G. *Org. Biomol. Chem.* **2018**, *16*, 1480–1488.
66. Hoover, J. M.; Stahl, S. S. *J. Am. Chem. Soc.* **2011**, *33*, 16901–16911.
67. Prathap, K. J.; Maayan, G. *Chem. Comm.* **2015**, *51*(55), 11096–11099.
68. Mohan, D. C.; Sadhukha, A.; Maayan, G. *J. Catal.* **2017**, *355*, 139–144.
69. Tian, H.; Yu, X.; Li, Q.; Wang, J.; Xu, Q. *Adv. Synth. Catal.* **2012**, *354*, 2671–2677.
70. Stamatin, Y.; Maayan, G. *Eur. J. Org. Chem.* **2020**, 3147–3152.
71. Hunter, B. M.; Gray, H. B.; Müller, A. M. *Chem. Rev.* **2016**, *116*, 14120–14136.
72. Zhang, T.; Wang, C.; Liu, S.; Wang, J. L.; Lin, W. *J. Am. Chem. Soc.* **2014**, *136*, 273–281.
73. Barnett, S. M.; Goldberg, K. I.; Mayer, J. M. *Nat. Chem.* **2012**, *4*, 498–502.

Index